工业和信息化
精品系列教材

uni-app
移动应用开发

（微课版）

赵丙秀 江骏 / 主编
龚丽 黄涛 于海平 / 副主编
罗保山 / 主审

Uni-app Mobile Application Development

人民邮电出版社
北京

图书在版编目（CIP）数据

uni-app 移动应用开发：微课版 / 赵丙秀，江骏主编. -- 北京：人民邮电出版社，2024.8. --（工业和信息化精品系列教材）. -- ISBN 978-7-115-64667-5

Ⅰ.TN929.53

中国国家版本馆 CIP 数据核字第 2024WY4613 号

内 容 提 要

  本书是一本系统介绍热门前端多平台框架 uni-app 的实用教程。全书共 7 章，主要包括两个部分。第一部分（第 1~4 章），内容包括初识 uni-app、uni-app 基础内容、uni-app 基础扩展、uni-app 组件等，每章都有综合案例，展示如何在实际应用场景中运用本章知识点。第二部分（第 5~7 章），内容包括 uni-app 的常用 API 和 3 个综合案例：智云翻译、仿网易云音乐 App 的音乐播放器、智慧环保项目。

  本书可以作为高等学校计算机相关专业移动应用开发、小程序开发课程的教材，也可以作为广大移动应用开发爱好者的自学用书。

◆ 主　编　赵丙秀　江　骏
　副主编　龚　丽　黄　涛　于海平
　主　审　罗保山
　责任编辑　刘　尉
　责任印制　王　郁　焦志炜

◆ 人民邮电出版社出版发行　北京市丰台区成寿寺路 11 号
　邮编　100164　电子邮件　315@ptpress.com.cn
　网址　https://www.ptpress.com.cn
　大厂回族自治县聚鑫印刷有限责任公司印刷

◆ 开本：787×1092　1/16
　印张：19.75　　　　　　　2024 年 8 月第 1 版
　字数：447 千字　　　　　　2025 年 1 月河北第 2 次印刷

定价：69.80 元

读者服务热线：(010)81055256　印装质量热线：(010)81055316
反盗版热线：(010)81055315
广告经营许可证：京东市监广登字 20170147 号

前　言

党的二十大报告指出，"我们要坚持教育优先发展、科技自立自强、人才引领驱动，加快建设教育强国、科技强国、人才强国，坚持为党育人、为国育才，全面提高人才自主培养质量，着力造就拔尖创新人才，聚天下英才而用之"。本书全面贯彻党的二十大精神，注重对学生社会主义核心价值观的培养，紧跟时代步伐，把握学生发展规律和实际特点，开拓进取、守正创新。

本书积极响应国家号召，贯彻落实教育部印发的《高等学校课程思政建设指导纲要》精神，将素养教育有机融入案例中。本书深入挖掘专业知识中蕴含的和可关联的思想价值和精神内涵，不仅嵌入中华优秀传统文化和社会主义建设成就等内容，还嵌入诚信教育、安全教育等内容，力图对本书的广度、深度进行拓展，增强本书的知识性、人文性。

移动应用包括 iOS、Android、HarmonyOS 等平台下的 App，以及各种应用（微信、支付宝、百度、今日头条、QQ、淘宝、钉钉）下的小程序、快应用。移动应用在用户规模及商业化方面都取得了极大的成功，针对不同的平台可以选择不同的开发环境和开发规范。同一应用如果需要多平台运行，需开发多套代码。DCloud 公司在 2018 年推出了基于 Vue.js 技术的开发所有前端应用的开源框架 uni-app，为开发者抹平了各平台的差异，实现了编写一套代码就可以发行到多个平台的目标。uni-app 凭借强大的跨平台能力，成为跨平台开发的首选框架。

本书编写团队具有多年移动应用开发经验、多门前端课程的教学经验，以及指导学生参加移动互联创新大赛的经验，本着"让读者容易上手、在实践中学会 uni-app 开发"的总体思路编写本书。本书简明易懂、循序渐进，实例丰富实用，知识点结合具体实例进行讲解，每章有综合案例、实训项目。

全书共 7 章，主要包括两个部分：第一部分（第 1～4 章）和第二部分（第 5～7 章）。主要内容如下。

第 1 章介绍 uni-app 发展历程、第 1 个 uni-app 项目、uni-app 打包和发行等内容。

第 2 章介绍 uni-app 的全局文件 pages.json、资源引用、页面样式、尺寸单位、基础组件、flex 布局等内容。

第 3 章介绍 uni-app 的生命周期、条件编译、扩展组件 uni-ui 等内容。

第 4 章介绍 uni-app 的组件，包括容器组件、基础组件、表单组件、媒体组件、地图组件等内容。

第 5 章介绍 uni-app 的常用 API，包括计时器、界面交互、网络、数据缓存、路由等内容。

第 6 章介绍 uni-app 的常用 API，包括媒体控制、文件操作、设备操作、登录等内容。

第 7 章详尽介绍一个综合案例——智慧环保项目。

学习本书需要具备 HTML5、CSS3、JavaScript、Sass、Vue.js 的相关知识。本书的 uni-app

项目基于 Vue.js 2 实现，对应的 Vue.js 3 的源代码参见本书的配套资源。

本书由武汉软件工程职业学院赵丙秀教授、江骏副教授担任主编，武汉软件工程职业学院龚丽、武汉城市职业学院黄涛、武汉软件工程职业学院于海平担任副主编，武汉软件工程职业学院胡佳静、武汉梦软科技有限公司熊泉浪、中国电力工程顾问集团中南电力设计院有限公司张尧、武汉升望科技有限公司王坚担任参编。全书由赵丙秀统稿、武汉软件工程职业学院罗保山主审。在编写过程中，本书得到了编者所在单位领导和同事的帮助和大力支持。

本书配有课程标准、实施方案、教案、PPT、源代码（含实训项目）等教学资源。请读者到人邮教育社区（www.ryjiaoyu.com）下载使用。

本书在编写过程中，参考并引用了许多专家、学者的著作和论文，在文中未一一注明。在此谨向相关参考文献的作者表示衷心的谢意。由于编者的水平和经验有限，书中难免存在不足之处，敬请读者批评指正。编者邮箱：sonyxiu@163.com。

编者

2024 年 1 月

目 录

第 1 章
初识 uni-app ································· 1
本章导读 ··· 1
学习目标 ··· 1
知识思维导图 ······································· 2
1.1 uni-app 发展历程 ···························· 2
    1.1.1 uni-app 的由来 ························ 2
    1.1.2 uni-app 的特点 ························ 3
1.2 第 1 个 uni-app 项目 ······················· 5
    1.2.1 uni-app 开发工具 ····················· 5
    1.2.2 新建项目 ································ 6
    1.2.3 项目目录结构 ························· 8
    1.2.4 运行项目 ······························ 10
1.3 uni-app 打包和发行 ······················ 14
    1.3.1 打包原生 App（云端）············ 14
    1.3.2 打包原生 App（本地）············ 15
    1.3.3 发行为 H5 ···························· 16
    1.3.4 发行为微信小程序 ················ 17
1.4 案例：示例代码 ···························· 17
本章小结 ·· 19
项目实战 ·· 19
拓展实训项目 ····································· 19

第 2 章
uni-app 基础内容 ·························· 20
本章导读 ·· 20
学习目标 ·· 20

知识思维导图 ····································· 21
2.1 pages.json ···································· 21
    2.1.1 globalStyle 和 pages ············· 21
    2.1.2 tabBar ································ 22
    2.1.3 subPackages ······················· 26
2.2 资源引用 ······································ 26
2.3 页面样式 ······································ 28
2.4 尺寸单位 ······································ 29
2.5 基础组件 ······································ 29
    2.5.1 view ··································· 30
    2.5.2 text ···································· 31
    2.5.3 navigator ···························· 33
    2.5.4 image ································ 34
    2.5.5 属性绑定和事件绑定············· 35
    2.5.6 v-for 渲染数据ยา····················· 37
2.6 flex 布局 ······································· 38
    2.6.1 flex 布局的基本概念ยา············· 38
    2.6.2 容器属性 ····························· 41
    2.6.3 子元素属性 ························· 55
2.7 案例：页面导航 ···························· 60
本章小结 ·· 64
项目实战 ·· 64
拓展实训项目 ····································· 65

第 3 章
uni-app 基础扩展 ·························· 66
本章导读 ·· 66
学习目标 ·· 66

知识思维导图	67
3.1 生命周期	67
3.1.1 应用生命周期	67
3.1.2 页面生命周期	68
3.1.3 组件生命周期	72
3.2 条件编译	72
3.3 扩展组件 uni-ui	75
3.3.1 安装组件	76
3.3.2 uni-scss 辅助样式	80
3.3.3 使用组件	84
3.4 案例：新闻列表页和新闻详情页	87
本章小结	90
项目实战	90
拓展实训项目	90

第 4 章

uni-app 组件 91

本章导读	91
学习目标	91
知识思维导图	92
4.1 容器组件	92
4.1.1 scroll-view	92
4.1.2 swiper	96
4.2 基础组件	101
4.2.1 rich-text	101
4.2.2 progress	103
4.3 表单组件	105
4.3.1 button	105
4.3.2 picker	109
4.3.3 slider	114
4.3.4 input	116
4.3.5 radio 和 checkbox	121

4.3.6 switch	124
4.3.7 textarea	126
4.3.8 form	129
4.4 媒体组件	132
4.4.1 camera	132
4.4.2 video	133
4.5 地图组件	137
4.6 案例一：典型注册页	138
4.7 案例二：典型个人中心页	141
本章小结	144
项目实战	145
拓展实训项目	145

第 5 章

常用 API（1） 146

本章导读	146
学习目标	146
知识思维导图	147
5.1 API 概述	147
5.2 计时器	148
5.2.1 设置计时器	148
5.2.2 取消计时器	149
5.3 界面交互	152
5.3.1 消息提示框	152
5.3.2 loading 提示框	154
5.3.3 模态框	155
5.3.4 操作菜单	156
5.4 网络	158
5.4.1 发起网络请求	158
5.4.2 上传文件	162
5.5 数据缓存	165
5.5.1 将数据缓存到本地	165

5.5.2　获取本地缓存数据……………167
　　5.5.3　清理本地缓存数据……………169
5.6　路由………………………………………174
　　5.6.1　路由 API……………………174
　　5.6.2　数据传递……………………177
5.7　案例：智云翻译…………………………178
本章小结…………………………………………192
项目实战…………………………………………192
拓展实训项目……………………………………192

第 6 章

常用 API（2）……………………………193

本章导读…………………………………………193
学习目标…………………………………………193
知识思维导图……………………………………194
6.1　媒体控制…………………………………194
　　6.1.1　audio 组件控制………………194
　　6.1.2　录音控制………………………202
　　6.1.3　图片控制………………………208
　　6.1.4　video 组件控制………………212
　　6.1.5　camera 组件控制……………216
6.2　文件操作…………………………………218
　　6.2.1　文件保存………………………218
　　6.2.2　文件选择………………………219
6.3　设备操作…………………………………219
　　6.3.1　系统 API………………………220
　　6.3.2　扫码……………………………221
　　6.3.3　拨打电话………………………223
　　6.3.4　剪贴板…………………………224
6.4　登录………………………………………225
6.5　案例：仿网易云音乐 App 的音乐
　　　播放器…………………………………227

本章小结…………………………………………247
项目实战…………………………………………248
拓展实训项目……………………………………248

第 7 章

智慧环保项目……………………………249

本章导读…………………………………………249
学习目标…………………………………………249
知识思维导图……………………………………250
7.1　项目介绍…………………………………250
　　7.1.1　项目概述………………………250
　　7.1.2　项目效果………………………251
7.2　环境配置…………………………………254
　　7.2.1　安装 json-server………………254
　　7.2.2　配置 json-server………………255
7.3　项目开发…………………………………258
　　7.3.1　创建初始项目…………………258
　　7.3.2　首页……………………………261
　　7.3.3　回收分类页、分类查询
　　　　　结果页……………………………266
　　7.3.4　注册页、登录页、个人
　　　　　中心页……………………………273
　　7.3.5　公司回收页、公司详情页、
　　　　　公司搜索结果页…………………284
　　7.3.6　下单页、订单页、订单
　　　　　详情页……………………………291
本章小结…………………………………………306
项目实战…………………………………………307
拓展实训项目……………………………………307

参考文献…………………………………………308

第1章
初识uni-app

本章导读

本章主要讲解 uni-app 的发展历程,以及 uni-app 开发工具、项目新建、项目目录结构、项目运行、项目打包和发行等内容。通过对本章的学习,读者可掌握 uni-app 项目的基本搭建方法。

学习目标

知识目标	1. 了解 uni-app 的发展历程 2. 熟悉 uni-app 项目的开发工具 3. 掌握 uni-app 项目的搭建方法 4. 熟悉 uni-app 项目的目录结构
能力目标	1. 能够熟练使用 HBuilderX 开发工具 2. 能够创建并运行 uni-app 项目 3. 能够熟悉 uni-app 项目的目录结构 4. 能够实现 uni-app 项目打包和发行
素质目标	1. 具有良好的软件编码规范素养 2. 培养技能报国的爱国主义情怀、精益求精的工匠精神 3. 激发对 uni-app 的学习兴趣

知识思维导图

1.1 uni-app 发展历程

uni-app 是一个多平台开发框架，开发者编写一套代码，可发行到 iOS、Android、Web（响应式）等的 App，以及各种小程序（微信、支付宝、百度、头条、飞书、QQ、快手、钉钉、淘宝等的小程序）、快应用等之上。

1.1.1 uni-app 的由来

很多人是在微信先接触小程序的，其实，DCloud 公司才是小程序领域的开创者。

DCloud 于 2012 年开始研发小程序技术，优化 WebView 的功能和性能，并加入 W3C 和 HTML5（简称 H5）中国产业联盟，推出了 HBuilder 开发工具，为后续产业化做准备。2015 年，DCloud 正式商用了自己的小程序，产品名为"流应用"。它不是 B/S 模式的轻应用，而是能接近原生功能的 App，并且即点即用，第一次使用时可以做到边下载边使用。为将小程序技术发扬光大，DCloud 将该技术标准捐献给工业和信息化部旗下的 HTML5 中国产业联盟，并推进微信和支付宝等接入该标准，以开展小程序业务。360 手机助手率先接入该标准，在其 3.4 版本中实现应用的"秒开"运行，如图 1-1 所示。

随后 DCloud 推动大众点评、携程旅行、京东、有道词典、唯品会等众多开发者为流应用平台提供应用。2015 年 9 月，DCloud 协助微信团队开展小程序业务，为其演示了流应用的"秒开"应用、扫码获取应用、分享链接获取应用等众多场景案例，以及分享了 WebView 体验优化的经验。微信团队经过分析，于 2016 年年初决定上线小程序业务，但其没有接入联盟标准，而是制定了自己的标准。

图 1-1　360 手机助手小程序

DCloud 持续在业内普及小程序理念，推进各互联网公司与手机厂商陆续上线类似小程序、快应用等业务。部分公司接入了联盟标准，但更多公司因纷争严重，标准难以统一。让开发者面对如此多的私有标准不是一件正确的事情，造成混乱的局面非 DCloud 所愿。于是 DCloud 决定开发一个免费开源的框架。既然各公司无法在标准上达成一致，那么就可以通过这个框架为开发者抹平各平台的差异。这就是 uni-app 的由来。uni-app 成功的因素有以下几点：

（1）经过多年积累，截至 2021 年 3 月，DCloud 已拥有 900 多万开发者；

（2）DCloud 一直都有小程序的 iOS、Android 引擎，因此 uni-app 的 App 平台和小程序平台保持高度一致；

（3）DCloud 在引擎上持续投入，因此 uni-app 的 App 平台功能、性能比大多数小程序平台的优秀；

（4）DCloud 对各平台小程序很了解，因此 uni-app 能成为抹平各平台差异的跨平台框架。

现在，uni-app 已经是业内最流行的应用框架之一，支撑着具有约 12 亿活跃手机用户的庞大生态。

1.1.2　uni-app 的特点

uni-app 是使用 Vue.js 开发跨平台应用的前端框架，可用于开发兼容多平台的应用，其特点如下。

1. 平台能力不受限

在跨平台的同时，uni-app 通过条件编译和平台特有的 API 调用，可以优雅地为某平台写个性化代码，调用专有功能而不影响其他平台。uni-app 支持原生代码混写和原生 SDK 集成。图 1-2 所示为 uni-app 功能框架，可以看出，uni-app 在跨平台的过程中不牺牲平台特色，可优雅地调用平台专有功能，真正做到海纳百川、各取所长。

图1-2　uni-app 功能框架

2. 一套代码运行到多个平台

uni-app 可实现一套代码同时运行到多个平台。如图 1-3 所示，一套代码同时运行到 iOS 模拟器、Android 模拟器、Web 模拟器、微信小程序、支付宝小程序、百度小程序、字节跳动小程序、QQ 小程序等，运行效果如图 1-4 所示。

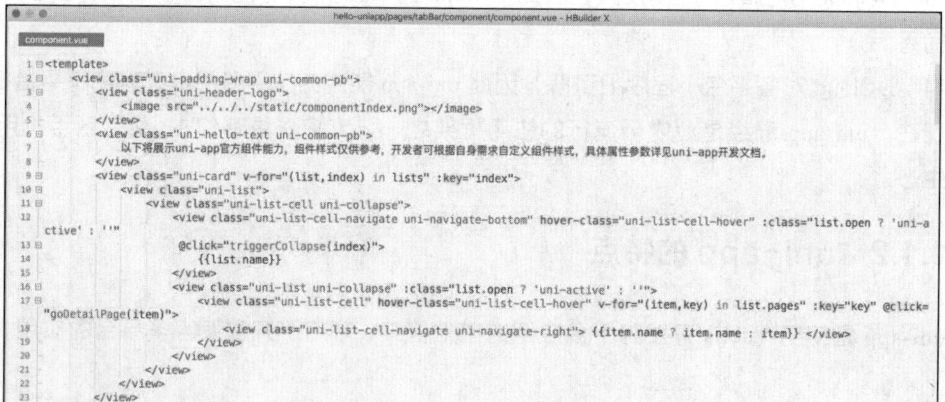

图1-3　一套代码运行到多个平台

图 1-4　实际运行效果

3. 性能体验优秀

uni-app 加载新页面的速度快，自动差量更新数据。其 App 平台支持原生渲染，可提供流畅的用户体验；小程序平台的性能优于市场其他框架。

4. 周边生态丰富

uni-app 支持通过 npm 安装第三方包，支持微信小程序组件和 SDK，微信生态的各种 SDK 可直接用于跨平台 App 开发，其组件市场有数千款组件。

5. 学习及开发成本低

uni-app 基于通用的前端技术栈，采用 Vue 语法和微信小程序 API，无额外学习成本，不仅开发成本低，招聘、管理、测试等各方面成本都较低。HBuilderX 是高效开发神器，熟练掌握后可使研发效率至少翻倍（即便只开发一个平台）。

1.2　第 1 个 uni-app 项目

1.2.1　uni-app 开发工具

uni-app 项目目录

搭建开发环境是使用 uni-app 的基础，下面先来看看 uni-app 开发工具。

目前前端界的主流 uni-app 开发工具有 4 个，分别是 Visual Studio Code、HBuilderX、WebStorm、Sublime Text。HBuilderX 和 uni-app 同属 DCloud 公司出品，HBuilderX 团队为 uni-app 做了大量的优化和定制。在 uni-app 日常开发中，使用比较多的是 HBuilderX

编辑器。HBuilderX 中的 H 是 HTML 的首字母，Builder 表示构造者，X 表示 HBuilder 的下一代版本。它也简称 HX，是轻如编辑器、强如 IDE 的合体版本。

HBuilderX 编辑器具有以下特点。

（1）HBuilderX 绿色发行包只有 10MB，很轻巧。

（2）不管是启动、打开大文档，还是编码提示，HBuilderX 都极速响应。它采用 C++的架构，性能远超 Java 或 Electron 架构。

（3）HBuilderX 对 Vue 做了大量优化，其开发体验远超其他开发工具。

（4）国外开发工具没有对我国的小程序开发进行优化，HBuilderX 可新建小程序等项目，为我国开发者提供高效的开发工具。

（5）HBuilderX 是唯一一个新建文件默认类型是 markdown 的编辑器，也是对 markdown 支持最强的编辑器之一，为 markdown 强化了众多功能。

（6）HBuilderX 的界面比其他工具的界面更清爽简洁，其绿柔主题界面经过科学的脑疲劳测试，是适合人眼长期观看的主题界面。

（7）现代 JavaScript（简称 JS）开发中有大量 JSON 结构的写法，HBuilderX 为其提供了比其他工具更高效的操作。

（8）HBuilderX 支持 Java 插件、Node.js 插件，并兼容很多 Visual Studio Code 的插件及代码块，可以通过外部命令方便地调用各种命令行功能，并设置快捷键。如果用户习惯使用其他工具的快捷键，可以选择【工具】-【预设快捷键方案切换】进行快捷键的切换。

HBuilderX 采用可视化的安装方式，安装比较简单。官方 IDE 下载界面如图 1-5 所示。

图 1-5　HBuilderX 官方 IDE 下载界面

单击【Download for Windows】按钮可下载最新的 Windows 平台 HBuilderX 安装包。单击【more】，可以选择下载其他平台或者历史版本的安装包。自 HBuilderX 3.4.6 起，取消发布单独的 App 开发版安装包，统一发布一个标准安装包。标准版 HBuilderX 也可以安装 App 相关插件。历史版本 3.4.6 之前的版本的安装包为 App 开发版安装包，一般比较大，在 300MB 左右。如果安装的是 App 开发版的 HBuilderX，则不需要安装其他插件，可以直接使用；如果安装的是标准版的 HBuilderX，在运行或发行 uni-app 时会提示安装 uni-app 插件，该插件安装完成后才能使用 HBuilderX。本书所用版本为 HBuilderX 3.6.13。

1.2.2　新建项目

创建项目是开发的第一步，uni-app 支持通过可视化界面、vue-cli 命令行两种方式快速创建项目。可视化界面方式比较简单，HBuilderX 内置相关环境，开箱即用，无须配置 Node.js。

实现步骤

（1）选择【文件】-【新建】-【项目】新建项目，如图 1-6 所示。

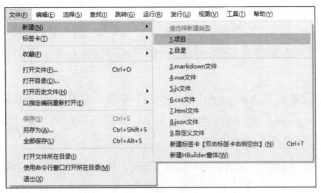

图 1-6　新建项目

（2）在弹出的【新建 uni-app 项目】对话框中选择 uni-app 类型，输入项目名称，并单击【浏览】按钮，选择项目存放地址；选择相应的 uni-app 项目模板，单击【创建】按钮即可，如图 1-7 所示。

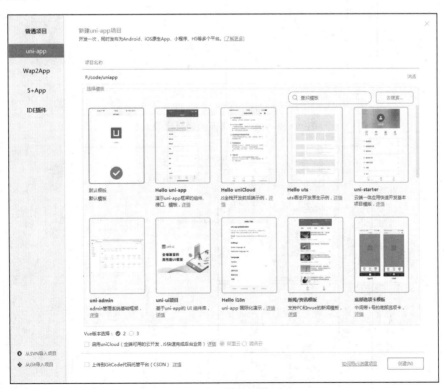

图 1-7　使用模板创建项目

初次创建项目，可以选择【默认模板】，这是 uni-app 默认的项目模板。在 uni-app 自带的模板中有 Hello uni-app 模板,它是官方的组件和 API 示例模板;还有一个重要模板是 uni-ui,它内置大量常用组件,日常开发时推荐使用该模板。

1.2.3 项目目录结构

在创建完 uni-app 项目后,系统会默认生成一些文件。这些文件就构成了初始项目目录。选择不同的模板生成的项目目录是不一样的。在日常开发过程中,需要在这个目录结构的基础上完成项目的开发。

uni-app 默认项目模板是没有 components 文件夹的,主要包含 pages、static 文件夹以及 App.vue、main.js、pages.json 等文件。如果使用 uni-ui 项目模板或其他项目模板,components 文件夹自动生成,相关新增的组件就会放到 components 文件夹里。

下面以常见的 uni-ui 模板、Hello uni-app 模板为例,讲解项目主要目录结构,由其他模板生成的目录结构与此类似。图 1-8 所示为 uni-ui 模板的项目目录结构,图 1-9 所示为 Hello uni-app 模板的项目目录结构。

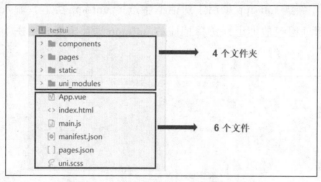

图 1-8　uni-ui 模板的项目目录结构

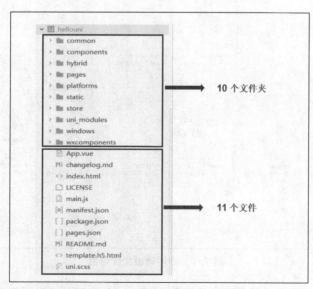

图 1-9　Hello uni-app 模板的项目目录结构

一个 uni-app 项目默认包含如下文件夹及文件（官方说明）。

```
├─uniCloud              云空间目录，例如阿里云的云空间目录为 uniCloud-aliyun
├─components            符合 Vue 组件规范的 uni-app 组件目录
│  └─comp-a.vue         可复用的 a 组件
├─utssdk                存放 UTS 文件的目录
├─pages                 业务页面文件存放的目录
│  ├─index
│  │  └─index.vue       index 页面文件
│  └─list
│     └─list.vue        list 页面文件
├─static                存放应用引用的本地静态资源（如图片、视频等）的目录
├─uni_modules           存放[uni_module](/uni_modules)的目录
├─platforms             存放各平台专用页面的目录
├─nativeplugins         存放 App 平台原生语言插件的目录
├─nativeResources       App 平台原生资源的目录
│  └─android            存放 Android 平台原生资源的目录
├─hybrid                App 平台存放本地 HTML 文件的目录
├─wxcomponents          存放小程序组件的目录
├─unpackage             非项目代码目录，一般存放运行或发行的编译结果
├─AndroidManifest.xml   Android 平台原生应用清单文件
├─main.js               Vue 初始化入口文件
├─App.vue               全局配置文件，用来配置 App 全局样式以及监听应用生命周期
├─manifest.json         配置应用名称、AppID、Logo、版本等打包信息
├─pages.json            配置页面路由、导航栏、选项卡等页面类信息
└─uni.scss              包含 uni-app 内置的常用样式变量
```

其中主要的文件及文件夹如表 1-1 所示。

表 1-1 主要的文件及文件夹

文件及文件夹	说明
common	存放公共 JS 和 CSS 文件的文件夹
components	存放组件的文件夹
pages	存放所有业务页面文件的文件夹
static	存放本地静态资源的文件夹
platforms	存放各平台专用页面的文件夹
unpackage	打包目录，新建项目是没有的
App.vue	项目的全局配置文件，用来配置 App 全局样式以及监听应用生命周期
main.js	Vue 初始化入口文件

续表

文件及文件夹	说明
manifest.json	应用配置文件，用于配置应用名称、AppID、Logo、版本等打包信息
pages.json	全局配置文件，用于配置页面路由、导航栏、选项卡等页面类信息
uni.scss	uni-app 的样式包，用于整体控制应用的风格

uni-app 还有一个特别之处，就是支持在 App、微信小程序等不同平台中使用其自定义的组件，其存放目录如表 1-2 所示。

表 1-2 跨平台自定义组件存放目录

平台	组件存放目录
App	wxcomponents
H5	wxcomponents
微信小程序	wxcomponents
支付宝小程序	mycomponents
百度小程序	swancomponents
字节跳动小程序	ttcomponents
QQ 小程序	wxcomponents

注意：项目目录结构不是一成不变的，我们可以按需使用，也可以继续在这个目录的基础上新建文件夹。

1.2.4 运行项目

项目创建完成后即可运行。uni-app 项目可以在多个平台上运行。下面以使用默认模板创建的 uni-app 项目为例来运行项目。

uni-app
项目运行

1. 在浏览器里运行

选择【运行】-【运行到浏览器】，在弹出的子菜单中选择需要的浏览器，这里选择 Chrome 浏览器，如图 1-10 所示。

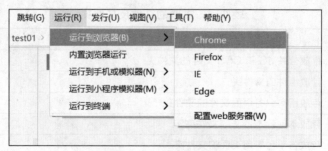

图 1-10 运行到浏览器

运行效果如图 1-11 所示。

2. 在真机里运行

通过 USB 将手机连接到计算机，开启手机的 USB 调试模式，选择【运行】—【运行到手机或模拟器】，在弹出的子菜单中选择需要的设备，如图 1-12 所示。如果无法识别手机，可以选择【真机运行常见故障排除指南】。

图 1-11　在浏览器中的运行效果　　　图 1-12　运行到真机

3. 在微信开发者工具里运行

选择【运行】—【运行到小程序模拟器】，在弹出的子菜单中选择【微信开发者工具】，如图 1-13 所示。

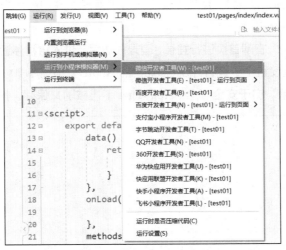

图 1-13　运行到微信开发者工具

说明：进行【运行到小程序模拟器】—【微信开发者工具】操作之前，必须搭建微信小程序环境，具体步骤如下。

（1）申请小程序账号

访问"微信公众平台"官网，单击【立即注册】按钮，打开注册页面，选择【小程序】，进入小程序注册页面，如图1-14所示。

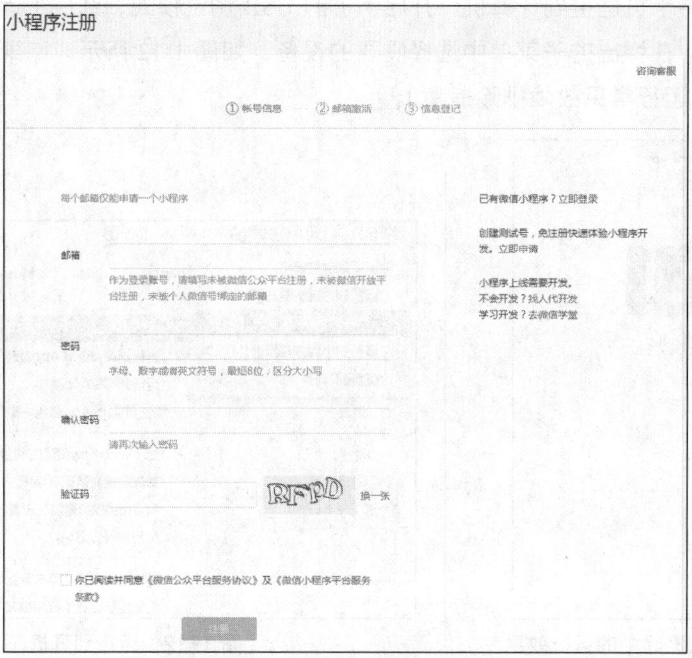

图1-14 小程序注册页面

根据提示填写信息，填写完成后单击下方的【注册】按钮，提交成功后，进入邮箱激活页面。打开注册邮箱，查看腾讯官方发送的小程序账号激活邮件，单击激活链接，继续注册流程。选择注册主体为"个人"，然后提交个人身份信息，并用微信扫码确认。认证成功后，即可拥有自己的微信小程序账号。

在"微信公众平台"官网首页进行登录，登录成功后，进入小程序管理页面。在【开发】—【开发管理】—【开发设置】中查看开发者ID中的AppID（小程序ID）和AppSecret（小程序密钥），如图1-15所示。出于安全考虑，AppSecret不再被明文保存，按照提示要求进行保存。

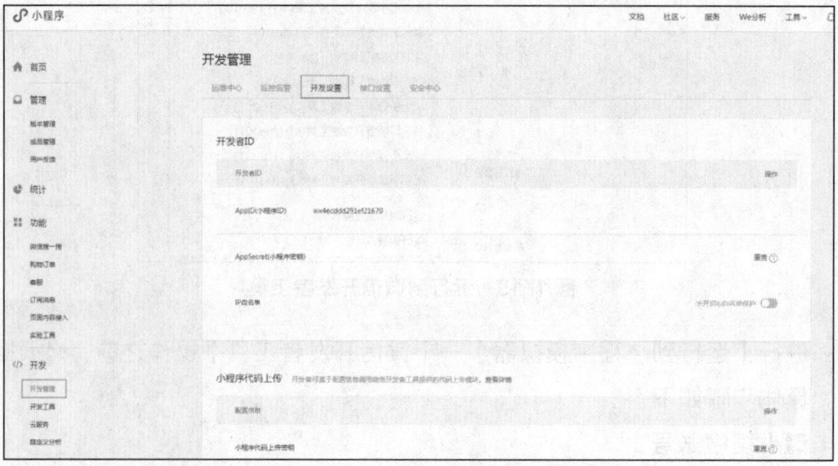

图1-15 小程序管理页面

通过微信小程序管理平台，可以管理小程序的权限、查看数据报表、发行小程序等。

（2）下载并安装微信开发者工具

在图 1-15 所示的页面中，选择【开发工具】—【开发者工具】，单击【下载】按钮，跳转到微信官方文档。选择【工具】—【下载】，打开微信开发者工具的下载页面，如图 1-16 所示。选择合适版本的安装包下载，并按照安装提示完成微信开发者工具的安装。

图 1-16　微信开发者工具的下载页面

（3）配置微信开发者工具

打开微信开发者工具，用微信扫码登录，其界面如图 1-17 所示。

图 1-17　微信开发者工具界面

进入微信开发者工具后，选择【设置】—【安全设置】。打开【设置】窗口，将【安全】设置中的【服务端口】功能打开，如图 1-18 所示。

图 1-18　微信开发者工具安全设置

如果是第一次使用微信开发者工具运行 uni-app 项目,需要先配置小程序 IDE 的相关路径,这样才能成功运行。在 HBuilderX 中,选择【工具】—【设置】—【运行配置】,在【小程序运行配置】中设置微信开发者工具的安装路径,如图 1-19 所示。

图 1-19　微信开发者工具路径配置

4. 在其他开发者工具里运行

uni-app 项目也可以在支付宝小程序等其他开发者工具里运行,方法与在微信开发者工具里运行相似,不再重复介绍。

1.3　uni-app 打包和发行

项目开发的最后一步是对项目进行打包发行。可以使用 HBuilderX 的云打包功能,将 uni-app 项目打包成安装包文件。

1.3.1　打包原生 App(云端)

在 HBuilder X 中需要登录 HBuilder 账号,并且实名认证才可以使用云打包功能。在 HBuilderX 中,选择【发行】—【原生 App-云打包】。如果没有登录 HBuilder 账号,则会弹出登录界面。请按照提示进行登录或注册,然后进行实名认证。

在已经登录并实名认证的情况下，会打开相应的打包配置对话框，如图 1-20 所示。配置打包选项，完成后单击【打包】按钮，进行打包。打包日志如图 1-21 所示。

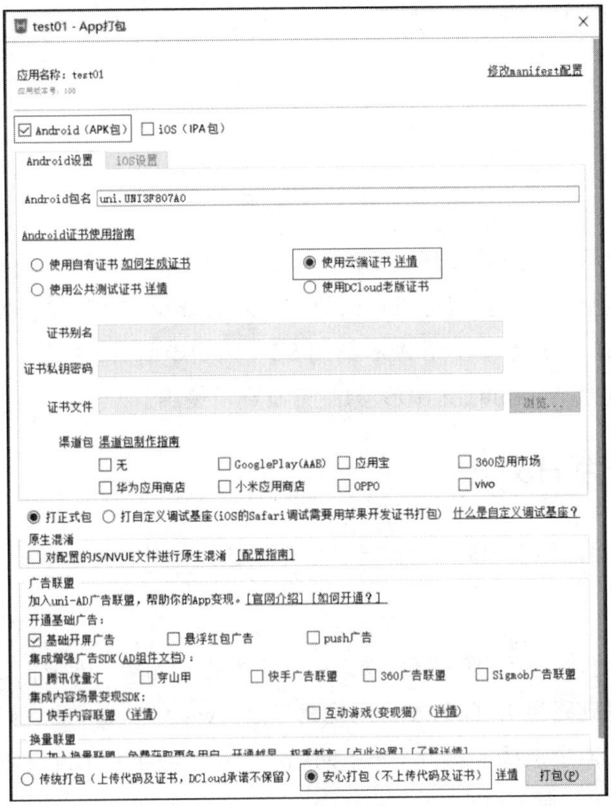

图 1-20　配置打包选项

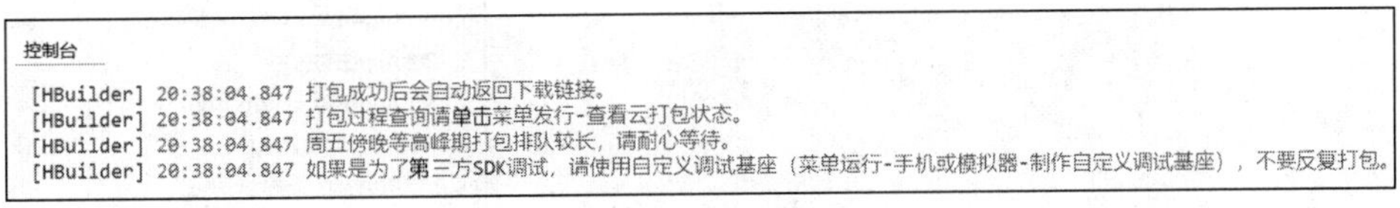

图 1-21　打包日志

打包成功后控制台会显示安装包文件地址，如图 1-22 所示。

图 1-22　打包成功提示

1.3.2　打包原生 App（本地）

如果要使用 Android Studio 进行本地打包，则选择 HBuilderX 的【发行】—【原生 App-本地打包】—【生成本地打包 App 资源】，如图 1-23 所示。

图 1-23　打包原生 App（本地）

App 本地打包要用到 App 离线开发工具包，即 App 离线 SDK，其把 App 运行环境（runtime）封装为原生开发调用接口，开发者可以在自己的 Android 及 iOS 原生开发环境中配置项目使用。从 HBuilderX 3.1.10 开始，使用 App 离线 SDK 需要申请 Appkey。

本地打包需要配合 Android 及 iOS 原生开发环境使用。

1.3.3　发行为 H5

如果 uni-app 项目需要在 H5 平台运行，则可以选择将其发行为 H5。

实现步骤

（1）在 manifest.json 的可视化界面中进行如下配置（如果 uni-app 项目发行在网站根目录下可不配置应用基础路径），此时发行网站路径是 www.***.com/h5，如图 1-24 所示。

（2）在 HBuilderX 中选择【发行】—【网站-PC Web 或手机 H5(仅适用于 uni-app)】，将 uni-app 项目发行为 H5，如图 1-25 所示。

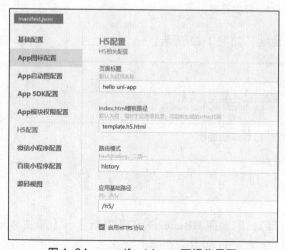

图 1-24　manifest.json 可视化界面

图 1-25　发行为 H5

1.3.4 发行为微信小程序

如果要将 uni-app 项目发行为微信小程序，可以选择【发行】—【小程序-微信(仅适用于 uni-app)】，uni-app 项目的代码会自动转换成小程序项目代码。

⚙ 实现步骤

（1）申请小程序账号并获取 AppID，相关方法在前面已介绍。在 HBuilderX 中选择【发行】—【小程序-微信(仅适用于 uni-app)】，在弹出的【微信小程序发行】对话框中对应的文本框内输入小程序名称或 AppID，如图 1-26 所示，单击【确定】按钮，即可在项目目录 unpackage/dist/build/mp-weixin 中生成微信小程序项目代码。

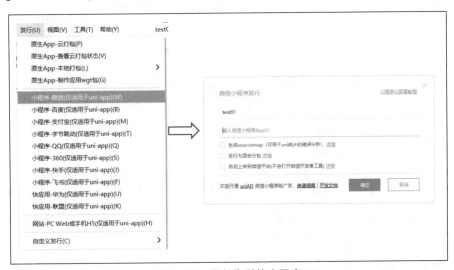

图 1-26　发行为微信小程序

（2）在微信开发者工具中，导入生成的微信小程序项目，测试项目代码运行正常后，单击【上传】按钮，之后按照提交审核、发行小程序的标准流程逐步操作即可。如果在【微信小程序发行】对话框中勾选了【自动上传到微信平台（不会打开微信开发者工具）】，则无须再打开微信开发者工具手动操作，将直接把项目上传到微信服务器提交审核。

1.4 案例：示例代码

使用 Hello uni-app 模板创建一个 uni-app 项目，本示例会展示 uni-app 的内置组件、扩展组件、接口的用法。

⚙ 实现步骤

（1）新建项目。选择【uni-app】—【Hello uni-app】，输入项目名"HelloTest_01"，如图 1-27 所示。

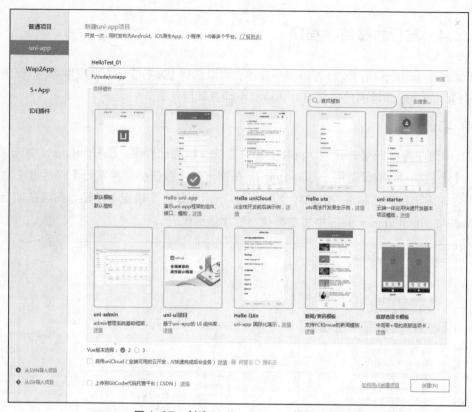

图 1-27　创建 Hello uni-app 模板项目

（2）熟悉项目目录结构，并将项目运行在浏览器、真机、微信开发者工具里，通过云打包生成文件。项目在浏览器中的运行效果如图 1-28 所示。

（a）内置组件页　　　　（b）接口页　　　　（c）扩展组件页　　　　（d）模板页

图 1-28　项目的运行效果

其中，图 1-28（a）展示了内置组件的使用方法，图 1-28（b）展示了接口的使用方法，图 1-28（c）展示了扩展组件 uni-ui 的各种组件的使用方法，图 1-28（d）展示了一些常用的模板，如顶部选项卡、列表到详情示例等。

本章小结

本章主要讲解了 uni-app 的发展历程，以及 uni-app 开发工具、项目新建、项目目录结构、项目运行、项目打包和发行等内容。通过对本章的学习，读者应掌握 uni-app 项目的基本搭建方法。

项目实战

使用 uni-app 默认模板创建项目，页面效果如图 1-29 所示。在该项目中会用到静态资源图片，需修改标题文字。

图 1-29　uni-app 项目页面效果

拓展实训项目

党的二十大报告指出"加快实现高水平科技自立自强"。移动应用开发包括微信、支付宝等各类平台小程序开发，以及 iOS、Android 等平台 App 开发等，请你搜索并了解它们的开发平台和技术，然后做一个简单的技术介绍页面，着重介绍国产技术。

第2章
uni-app基础内容

本章导读

本章主要讲解 uni-app 的配置文件 pages.json、资源引用、页面样式、尺寸单位、常用基础组件、flex 布局等内容。通过对本章的学习,读者能够掌握 uni-app 开发的基础内容。

学习目标

知识目标	1. 掌握常用的基础组件 2. 掌握 flex 布局
能力目标	1. 能够熟练使用基础组件搭建页面 2. 能够熟练使用 flex 布局 3. 能够开发多 tabBar 应用
素质目标	1. 具有良好的软件编码规范素养 2. 培养独立思考、分析问题、解决问题的能力 3. 具备持之以恒的精神

知识思维导图

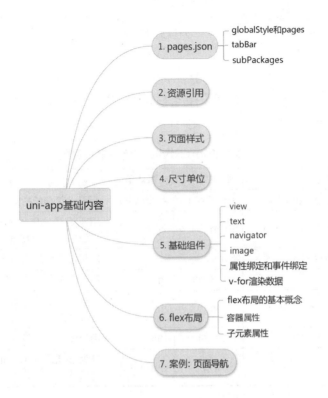

2.1 pages.json

pages.json 文件用来对 uni-app 进行全局配置,如设置页面文件的路径、窗口样式、原生导航栏、底部的原生 tabBar 等。

pages.json 文件中有 globalStyle、pages、easycom、tabBar、condition、subPackages 等 12 个配置项。本节讲解常用配置项 globalStyle、pages、tabBar、subPackages,其他的配置项请读者参考 uni-app 官网。

2.1.1 globalStyle 和 pages

在 HBuliderX 中新建一个默认模板项目 uniappch02,打开 pages.json,代码如下。

```
{
    "pages":[ //pages 数组中第一项表示应用启动页,参考 uni-app 官网
        {
            "path":"pages/index/index",
            "style":{
                "navigationBarTitleText":"uni-app"
            }
        }
```

```
        }
    ],
    "globalStyle":{
        "navigationBarTextStyle":"black",
        "navigationBarTitleText":"uni-app",
        "navigationBarBackgroundColor":"#F8F8F8",
        "backgroundColor":"#F8F8F8"
    }
}
```

globalStyle 节点用于全局设置应用的状态栏、导航栏、标题、窗口背景色等。globalStyle 节点的常用属性如表 2-1 所示。

表 2-1 globalStyle 节点的常用属性

属性	类型	默认值	描述
navigationBarBackgroundColor	HexColor	#F7F7F7	用于设置导航栏的背景颜色
navigationBarTextStyle	String	white	用于设置导航栏的标题颜色及状态栏颜色,仅支持 black/white
navigationBarTitleText	String		用于设置导航栏标题文字
backgroundColor	HexColor	#ffffff	用于设置下拉显示的窗口背景色
backgroundTextStyle	String	dark	用于设置下拉 loading 的样式,仅支持 dark/light
enablePullDownRefresh	Boolean	false	用于设置是否开启下拉刷新
navigationStyle	String	default	值为 default/custom,custom 用于取消默认原生导航栏

pages 节点用于配置应用由哪些页面组成,pages 节点接收一个数组,数组的每一项都是一个对象。其中第一项为应用启动页(首页),应用中新增或减少页面时,都需要对 pages 数组进行修改。pages 节点的常用属性如表 2-2 所示。

表 2-2 pages 节点的常用属性

属性	类型	描述
path	String	配置页面路径
style	Object	配置页面窗口表现,参考表 2-1,style 中的配置属性与 globalStyle 节点的常用属性一致

path 属性中的文件名不需要扩展名,框架会自动寻找路径下的页面资源。
style 属性中的配置项与 globalStyle 节点的配置项不同时,会覆盖 globalStyle 节点的配置项。

2.1.2 tabBar

tabBar 节点用于实现多页面的切换。对于一个多 tabBar 应用,可以通过 tabBar 节点配置项指定一级导航栏,以及 tabBar 切换时显示的对应页面。在 pages.json 中提供 tabBar 节点配置，不仅是为了方便快速开发导航，更重要的是提升 App 平台和小程序平台的性能。tabBar 节点的常用属性如表 2-3 所示。

表 2-3 tabBar 节点的常用属性

属性	类型	默认值	描述
color	HexColor		用于设置 tabBar 上的文字默认颜色，必填
selectedColor	HexColor		用于设置 tabBar 上的文字被选中时的颜色，必填
backgroundColor	HexColor		用于设置 tabBar 的背景颜色
borderStyle	String	black	用于设置 tabBar 上边框的颜色，可选值有 black/white
list	Array		用于设置 tabBar 的列表，最少有 2 个，最多有 5 个

list 接收一个数组，数组中每一项都是一个对象。list 数组中对象的属性如表 2-4 所示。

表 2-4 list 数组中对象的属性

属性	类型	描述
pagePath	String	用于设置页面路径，必须在 pages 节点中先定义，必填
text	String	用于设置 tabBar 上的按钮文字，在 App 和 H5 平台非必填
iconPath	String	用于设置图片路径，icon 大小限制为 40KB，建议尺寸为 81px×81px，当 position 为 top 时，此属性无效，不支持网络图片和字体图标
selectedIconPath	String	选中图片的路径，图片要求与 iconPath 的一致

【实例 2-1】实现一个有两个 tabBar 页面的应用。

在本例中，沿用前面的 uniappch02 项目，新建两个 tabBar 页面，分别为基础组件页和 flex 布局页。在内置浏览器上显示的效果如图 2-1 所示。

（a）基础组件页

（b）flex 布局页

图 2-1 uniappch02 项目的演示效果

实现步骤

（1）准备 tabBar 上的图标。打开 iconfont 官网，在搜索框中输入"组件"，如图 2-2 所示。

图 2-2　iconfont 官网首页

下载具有两种颜色（亮色、暗色）的、尺寸为 64px×64px 的 PNG 格式的同一种透明图片，如图 2-3 所示。再下载一组"flex"图标，然后将下载的图标复制到 static 文件夹。

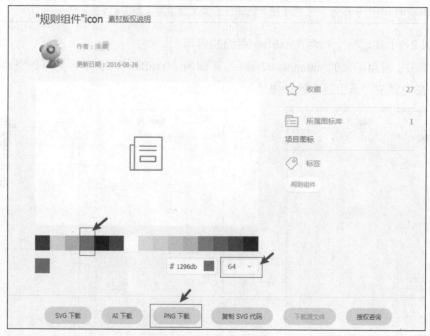

图 2-3　下载图片页面

（2）新建页面。在 pages 节点上右击，在弹出的列表中选择【新建页面】，打开【新建 uni-app 页面】窗口，如图 2-4 所示，输入文件名，单击【创建】按钮。

新建 compony.vue、flex.vue 文件。文件结构如图 2-5 所示。

图 2-4 【新建 uni-app 页面】窗口

图 2-5 文件结构

打开 compony.vue、flex.vue 文件，分别在<view></view>之间输入内容：基础组件演示、弹性盒子布局 flex。以下代码是 flex.vue 文件的代码，compony.vue 文件的代码同理。

```
<template>
    <view>   弹性盒子布局 flex   </view>
</template>
```

（3）修改 pages.json 文件。打开 pages.json 文件，可以看到在 pages 节点中多了两个页面。将 compony 页面移到第一项，并修改 navigationBarTitleText 属性，然后新增 tabBar 节点。

```
{
    "pages":[ //pages 数组中第一项表示应用启动页，参考 uni-app 官网
        {
            "path" :"pages/compony/compony",
            "style" :{
                "navigationBarTitleText":"基础组件",
                "enablePullDownRefresh":false
            }
        },
        {
            "path":"pages/index/index",
            "style":{
                "navigationBarTitleText":"uni-app"
            }
        },
        {
            "path" :"pages/flex/flex",
            "style" :{
                "navigationBarTitleText":"flex 布局",
                "enablePullDownRefresh":false
            }
```

```
        }
    ],
    "tabBar":{
        "backgroundColor":"#F8F8F8",
        "color":"#8F8F94",
        "list":[ {"text":"组件",
            "pagePath":"pages/compony/compony",
            "iconPath":"static/unComponet.png",
            "selectedIconPath":"static/componet.png"
            },
            {"text":"flex 布局",
            "pagePath":"pages/flex/flex",
            "iconPath":"static/unflex.png",
            "selectedIconPath":"static/flex.png"
            } ]
    },
    "globalStyle":{
        "navigationBarTextStyle":"black",
        "navigationBarTitleText":"uni-app",
        "navigationBarBackgroundColor":"#F8F8F8",
        "backgroundColor":"#F8F8F8"
    }
}
```

2.1.3 subPackages

subPackages 节点用于为小程序的分包加载配置。因小程序有体积和资源加载限制，各小程序平台提供了分包方式，以加快小程序的下载和启动速度。主包用于放置默认启动页面、babBar 页面，以及一些所有分包都会用到的公共资源或 JS 脚本；而分包则根据 pages.json 的配置进行划分。

在小程序启动时，会默认下载主包并启动主包内的页面，当用户打开分包内某个页面时，对应分包会被自动下载下来，下载完成后再进行展示。此时终端界面会有等待提示。

subPackages 节点接收一个数组 list，list 数组的每一项都是应用的分包，其元素的属性如表 2-5 所示。设置 subPackages 节点后，再设置 preloadRule 分包预加载配置，在打开小程序的某个页面时，由框架自动预加载可能需要的分包，加快打开后续分包页面的速度。

表 2-5 list 数组元素的属性

属性	类型	是否必填	描述
root	String	是	用于设置分包的根目录
pages	Array	是	用于设置分包由哪些页面组成

2.2 资源引用

在 Vue 页面中，会引用 JS 文件、CSS 文件或一些图片、视频等静态资源。

1. 引用 JS 文件

在 JS 文件或<script></script>标签内引用 JS 文件时，可以使用相对路径和绝对路径。绝对路径以"@"开头，"@"指向项目根目录，在 cli 项目中"@"指向 src 目录。JS 文件不支持使用"/"开头的方式引用。在相对路径中".."表示当前目录的上一级目录，"."代表当前所在目录。示例如下。

```
import add from '@/common/add.js';        // 绝对路径
import add from '../../common/add.js';    // 相对路径
```

uni-app 支持使用 npm 命令安装第三方包。使用 npm 命令安装完第三方包后，可以直接在<script></script>中引用它们。

【实例 2-2】在 VUE 文件中引用使用 npm 命令安装的包。

在本例中，安装 sha256 算法包，对文本信息进行加密。使用 npm 命令安装 js-sha256 包，然后在 index.vue 页面中引用并使用它。

实例 2-2

实现步骤

（1）在 HBuilderX 中选择【视图】—【显示终端】。如果不是 cli 项目，而是在 HBuilderX 中创建，则需要使用 npm 命令初始化 npm 项目。在终端输入以下代码。

```
npm init -y
```

注意：如果计算机上没有安装 Node.js，则可能无法识别 npm 命令。因此需要到 Node.js 官网下载并安装 Node.js。其官网首页如图 2-6 所示。

图 2-6　Node.js 官网首页

安装 Node.js 后，打开命令提示符窗口，输入"node -v"命令，可以查看 Node.js 的版本，如图 2-7 所示。

（2）安装 js-sha256 包。

```
npm install js-sha256 --save
```

修改 pages/index/index.vue 文件，如图 2-8 所示。

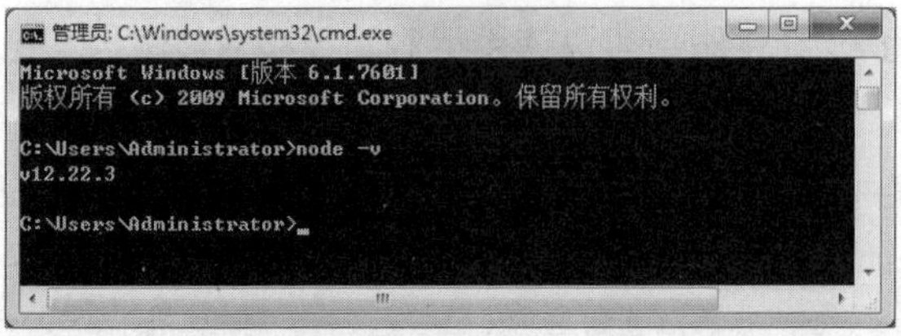

图 2-7 查看 Node.js 版本 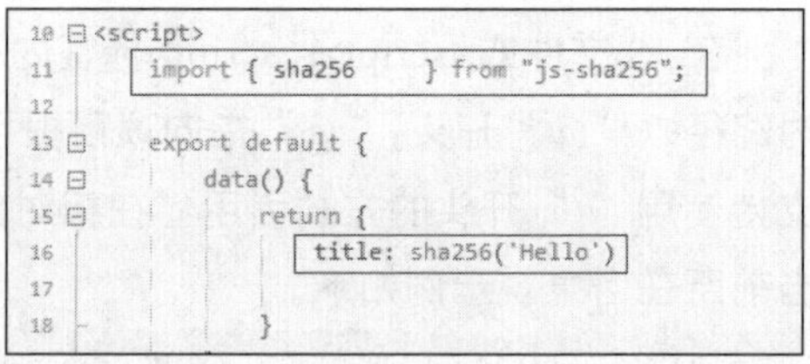　　　　图 2-8 修改 pages/index/index.vue 文件

（3）查看 index.vue 页面效果，可以看到页面中的"Hello"变为一串加密数字。

2. 引用 CSS 文件

在 uni-app 中，通过@import 语句导入 CSS 文件。从 1.4 节的 HelloTest_01 的示例代码中将 common 文件夹复制到项目 uniappch02 中，将 static/uni.ttf 文件复制到 uniappch02 的 static 文件夹中。打开 index.vue，在<style></style>标签内添加下列任意一行代码。

```
@import "../../common/uni.css";
@import url('/common/uni.css');
@import url('@/common/uni.css');
@import url('../../common/uni.css');
```

引用 LESS、SCSS 文件的方法类似。

3. 在<template></template>内引用静态资源

在<template></template>内可引用静态资源。在<image></image>、<video></video>等标签的 src 属性中，可以使用相对路径或者绝对路径。

```
<image class="logo" src="/static/logo.png"></image>
<image class="logo" src="@/static/logo.png"></image>
<image class="logo" src="../../static/logo.png"></image>
```

2.3　页面样式

在 uni-app 项目中，一个页面就是一个符合 Vue SFC 规范的 VUE 文件。

uni-app 的组件支持使用 style、class 属性来控制样式。静态样式统一写到 class 属性中，style 属性可接收动态样式，在运行时会对其进行解析。我们应尽量避免将静态样式写进 style 属性中，以免影响渲染速度。

uni-app 支持的选择器有 class、#id、element、element,element、:after、:before，不支持使用*选择器。在微信小程序自定义组件中仅支持 class 选择器。

page 节点相当于 HTML 中的 body 节点。以下代码设置的是页面的背景颜色。

```
page{
    background-color:#CCCCCC;
}
```

定义在 App.vue 中的样式为全局样式，作用于每一个页面。在 pages 目录下的 VUE 文件中定义的样式为局部样式，只作用于相应页面，并会覆盖 App.vue 中相同选择器的设置。

2.4 尺寸单位

uni-app 支持的通用 CSS 尺寸单位包括 px、rpx，其中 px 为屏幕像素，rpx 为响应式 px。

rpx 是一种根据屏幕宽度自适应的动态单位。以 750 物理像素宽的屏幕为基准，750rpx 恰好为屏幕宽度。屏幕变宽，rpx 实际显示效果会等比放大，但在 App 平台和 H5 平台，屏幕宽度达到 960px 时，默认将按照 375px 的屏幕宽度进行计算。在开发移动端项目时通常选择 rpx 作为尺寸单位。

设计师在提供设计图时，一般只提供一个分辨率的图。若严格按设计图标注的 px 进行开发，在不同宽度的手机上，界面很容易变形，而且主要会发生宽度变形。高度一般因为有滚动条，不容易出问题。由此，带来了较强的动态宽度单位需求。微信小程序使用 rpx 以解决这个问题。uni-app 在 App 平台、H5 平台都支持 rpx，并且可以配置不同屏幕宽度的计算方式。

页面元素宽度在 uni-app 中的计算公式：

750×元素在设计图中的宽度 / 设计图基准宽度

（1）若设计图基准宽度为 750px，元素 A 在设计图上的宽度为 100px，那么元素 A 在 uni-app 里面的宽度应该为：750×100 / 750，结果为 100rpx。

（2）若设计图基准宽度为 640px，元素 B 在设计图上的宽度为 100px，那么元素 B 在 uni-app 里面的宽度应该为：750×100 / 640，结果约为 117rpx。

（3）若设计图基准宽度为 375px，元素 C 在设计图上的宽度为 200px，那么元素 C 在 uni-app 里面的宽度应该为：750×200 / 375，结果为 400rpx。

2.5 基础组件

组件是视图层的基本组成单元。组件是单独且可复用的功能模块的封装。每个组件包括如下几个部分：以组件名称为标记的开始标签和结束标签、组件内容、组件属性、组件属性值。

组件名称由尖括号包裹，称为标签，它有开始标签和结束标签。结束标签的"<"后面用"/"来表示结束。结束标签也称为闭合标签。

开始标签和结束标签之间的部分称为组件内容。开始标签中可以设置属性，属性可以有多个，多个属性之间用空格分隔。每个属性通过"="赋值。例如下面的代码。

```
<image class="logo" src="/static/logo.png"></image>
```

每个 VUE 文件的根节点必须为<template></template>，且它下方只能且必须有一个 view 组件。

```
<template>
    <view>
            <text>弹性盒子布局 flex</text>
    </view>
</template>
```

uni-app 提供了一系列的基础组件，包括容器组件、表单组件、媒体组件等，本节主要讲解几个常用的组件。本节案例使用的页面为 compony.vue。

2.5.1 view

view 是容器组件，类似于 HTML 中的<div></div>标签，用于包裹各种元素内容，是页面布局常用的组件。view 组件的属性如表 2-6 所示。

表 2-6 view 组件的属性

属性	类型	默认值	说明
hover-class	String	none	指定按下去的样式类。当 hover-class="none" 时，没有点击态效果
hover-stop-propagation	Boolean	false	指定是否阻止某节点的祖先节点出现点击态
hover-start-time	Number	50	按住后多久出现点击态，单位为毫秒（ms）
hover-stay-time	Number	400	手指松开后点击态保留时间，单位为毫秒（ms）

【实例 2-3】演示 view 组件的属性用法。

在本实例中，添加一个 view 组件，通过设置 hover-class 属性的值，实现点击该 view 组件，改变其背景颜色。在该 view 组件中放置一个更小的 view 组件，通过 hover-stop-propagation 属性设置是否阻止冒泡行为。演示效果如图 2-9 所示。

实例 2-3

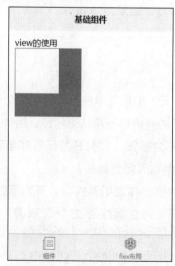

图 2-9 组件页演示效果（1）

具体代码如下。

```
<template>
<view class="content">
    <view class="viewbox">
        <text>view 的使用</text>
```

```
            <view class="outbox" hover-class="outhover" hover-start-time="500" hover-stay-time="1600">
                <view class="inbox" hover-class="inhover" hover-start-time="500" hover-stay-time="1600" :hover-stop-propagation="true"></view>
            </view>
        </view>
    </view>
</template>
<style>
.content {
        display:flex;
        flex-direction:column;
        align-items:flex-start;
        justify-content:center;
}
.viewbox{
        padding:30rpx;
}
.outbox{
        width:300rpx;
        height:300rpx;
        background-color:#DC7004;
        border:1px solid #DC7004;
}
.inbox{
        width:200rpx;
        height:200rpx;
        background-color:#FFFFFF;
}
.outhover{
        background-color:#007AFF;
}
.inhover{
        background-color:#0EA391;
}
</style>
```

2.5.2 text

text 组件用于包裹文本内容，类似于 HTML 中的 <span></span> 标签。其属性如表 2-7 所示。

表 2-7　text 组件属性

属性	类型	默认值	说明
selectable	Boolean	false	指定文本是否可选（在微信小程序中用 user-select 属性）
space	String		指定显示连续空格
decode	Boolean	false	指定是否解码

decode 可以解析的内容有 、<、>、&、'、 、 。space 属性值的说明如表 2-8 所示。

表 2-8 space 属性值的说明

值	说明
ensp	中文字符空格的一半大小
emsp	中文字符空格的大小
nbsp	根据字体设置的空格大小

【实例 2-4】演示 text 组件属性用法。

本例中文本内容含有<、>、空格。读者可以自行修改 space 属性的值，查看页面效果。页面效果如图 2-10 所示。

图 2-10 组件页演示效果（2）

具体代码如下。

（1）在 compony.vue 文件中视图层的<template></template>中添加以下代码：

```
<template>
    ……省略
    <view class="textbox">
            <text  space="ensp"  :decode="true" >{{words}}</text>
    </view>
    ……省略
</template>
```

（2）在<script></script>添加以下代码：

```
<script>
export default {
    data() {
        return {
            words:'永不放弃是实现梦想的唯一途径！&lt;语录&gt;'
        }
    },
    methods:{
    }
}
</script>
```

2.5.3 navigator

navigator 组件用于实现页面跳转，类似于 HTML 中<a></a>标签，但只能实现本地页面的跳转。注意，目标页面必须在 pages.json 中注册。该组件常用的属性如表 2-9 所示。

表 2-9 navigator 组件常用的属性

属性	类型	默认值	说明
url	String		指定应用内的跳转链接，值为相对或绝对路径
open-type	String	navigate	指定跳转方式，取值如表 2-10 所示
hover-class	String	navigate-hover	指定点击时的样式类，当值为 none 时，没有点击态

open-type 属性的有效值如表 2-10 所示。

表 2-10 open-type 属性的有效值

有效值	说明
navigate	保留当前页面，跳转到应用内的某个页面
redirect	关闭当前页面，跳转到应用内的某个页面
switchTab	跳转到 tabBar 页面，并关闭其他所有非 tabBar 页面
reLaunch	关闭所有页面，打开应用内的某个页面
navigateBack	关闭当前页面，返回上一页面或多级页面

【实例 2-5】演示 navigator 组件的 3 种不同形式的导航。

在本例中，3 个按钮可实现不同形式的跳转。点击第一个按钮跳转的新页面（这里设置为 index.vue）中有【返回】按钮。点击第二个按钮关闭当前页后跳转到其他页面（这里设置为 index.vue）。点击第三个按钮跳转到 tabBar 页面（这里设置为 flex.vue）。演示效果如图 2-11 所示。

实例 2-5

（a）组件页

（b）index.vue 页面

图 2-11 组件页和 index.vue 页面演示效果

在 compony.vue 文件中的视图层<template></template>里面添加以下代码：

```
<view class="navbox">
    <text>navigator 的使用</text>
    <view class="btnnav">
        <navigator url="/pages/index/index" >
            <button type="primary" size="mini">新页面</button>
        </navigator>
        <navigator url="/pages/index/index" open-type="redirect" >
            <button type="primary" size="mini">当前页</button>
        </navigator>
        <navigator url="/pages/flex/flex" open-type="switchTab" >
            <button type="primary" size="mini">tabBar 页</button>
        </navigator>
    </view>
</view>
```

在<style></style>里面添加对应的样式：

```
.navbox{
    width:100%;
}
.btnnav{
    display:flex;
    justify-content:space-between;
    margin-top:20rpx ;
}
```

2.5.4 image

image 组件用来展示图片，默认宽度为 320px、高度为 240px，其 src 属性支持相对路径、绝对路径，支持 base64 码。image 组件常用属性如表 2-11 所示。

表 2-11 image 组件常用属性

属性	类型	默认值	说明
src	String		指定图片资源地址
mode	String	'scaleToFill'	指定图片裁剪、缩放的模式

其中，mode 一共有 14 个有效值，包括 5 种缩放模式的值（见表 2-12）和 9 种裁剪模式的值。

表 2-12 mode 的缩放模式取值

值	说明
scaleToFill	不保持宽高比缩放图片，使图片的宽度完全拉伸至填满 image 组件
aspectFit	保持宽高比缩放图片，使长边完全显示出来
aspectFill	保持宽高比缩放图片，使短边完全显示出来，在长边方向将会发生截取
widthFix	宽度不变，高度自动变化，保持原图宽高比不变
heightFix	高度不变，宽度自动变化，保持原图宽高比不变

【实例 2-6】在 compony 页面添加一个 image 组件。演示效果如图 2-12 所示。

实例 2-6

图 2-12　组件页演示效果（3）

具体代码如下。

（1）在<template></template>部分添加以下代码：

```
<view class="imgbox">
    <image src="../../static/奋斗.jpg" mode="widthFix" class="img"></image>
</view>
```

（2）在<style></style>部分添加以下代码：

```
.imgbox{
    width:100%;
    text-align:center;
}
.img{
    width:60%;
}
```

2.5.5　属性绑定和事件绑定

uni-app 中渲染后端数据的方式和 Vue 的方式一样，直接用"{{ }}"包裹。例如实例 2-3 中的 text 组件。

```
<text space="ensp" :decode="true" >{{words}}</text>
```

其中，{{words}}中的 words 来自 Script 代码中的"words:'永不放弃是实现梦想的唯一途径！<语录>'"。

1. 属性绑定

组件的属性绑定不能直接使用{{}}，而要使用 v-bind。此时，v-bind 可以简写为":"。

```
<template>
    <view class="imgbox">
        <image :src="imgpath" mode="widthFix" class="img"></image>
    </view>
</template>


<script>
export default {
    data() {
        return {
            words:'永不放弃是实现梦想的唯一途径! &lt;语录&gt;',
            imgpath:'../../static/奋斗.jpg'
        }
    },
    methods:{
    }
}
</script>
```

2. 事件绑定

每个组件都有事件,触发事件就是在指定的条件下触发某个 JS 方法。比如点击按钮,就会触发@click 事件。事件也是组件的属性,只不过这类属性以@为前缀。事件的属性值为在 <script></script> 的 methods 里定义的 JS 方法。

【实例 2-7】在 image 组件下方添加一个按钮,点击按钮弹出提示框。演示效果如图 2-13 所示。

实例 2-7

图 2-13 组件页演示效果(4)

具体代码如下。

（1）<template></template>部分的代码：

```
<button @click="test()">点击试一下</button>
```

（2）<script></script>部分的代码：

```
<script>
export default {
    data() {
        return {
            words:'永不放弃是实现梦想的唯一途径！&lt;语录&gt;',
            imgpath:'../../static/奋斗.jpg'
        }
    },
    methods:{
        test(){
            uni.showToast({
                title:' 你刚才点击了按钮！',
                duration:1500
            })
        }
    }
}
</script>
```

2.5.6 v-for 渲染数据

在 uni-app 中，可以用 v-for 来循环遍历数据，这与 Vue 中是一样的。

【实例 2-8】演示使用 v-for 渲染数据。

在本例中，在组件页中显示 list 数组的内容。演示效果如图 2-14 所示。

实例 2-8

图 2-14　组件页演示效果（5）

具体代码如下。

```
<template>
    <view class="content">
        ……省略前面的代码
        <view class="listbox">
            <text v-for="(item,index) in list" :key="index">
                {{item.name}} -----{{item.person}}
            </text> </br>
        </view>
    </view>
</template>
<script>
    export default {
        data() {
            return {
                words:'永不放弃是实现梦想的唯一途径！&lt;语录&gt;',
                imgpath:'../../static/奋斗.jpg',
                list:[{
                    id:1,
                    name:'神舟十四号',
                    person:'陈冬、刘洋、蔡旭哲'
                },
                {
                    id:2,
                    name:'神舟十三号',
                    person:'翟志刚、王亚平、叶光富'
                }
                ]
            }
        },
……省略前面实例中的代码
        }
    }
</script>
```

2.6 flex 布局

2.6.1 flex 布局的基本概念

　　flex 布局是在 CSS3 中引入的，又称为"弹性盒子模型"。使用 flex 布局可以轻松地创建响应式网页布局。flex 布局改进了块模型，既不使用浮动，又不合并弹性盒子容器与其内容之间的外边距。它是一种非常灵活的布局方法，就像将几个小盒子放在一个大盒子里一样，它们相互独立，方便设置。

　　弹性盒子由容器、子元素和轴构成。在默认情况下，子元素的排布方向与主轴的方向是一致的，弹性盒子如图 2-15 所示。flex 布局可以用简单的方式满足很多常见的复杂布局需求，它的优势在于开发人员只需要声明布局应该具有的行为，而不需要给出具体的实现方式。

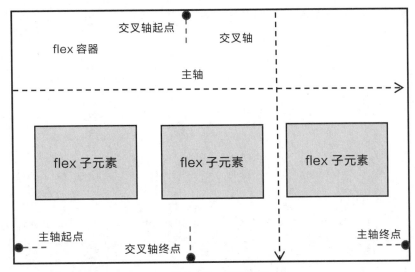

图 2-15　弹性盒子

在 flex 布局中，用于包含内容的组件称为容器，容器内部的组件称为项目或子元素。容器默认存在两根轴：水平的主轴和垂直的交叉轴（或侧轴）。主轴的开始位置（主轴与边框的交叉点）叫作主轴起点，结束位置叫作主轴终点；交叉轴的开始位置（交叉轴与边框的交叉点）叫作交叉轴起点，结束位置叫作交叉轴终点。

通过 view 组件设置 display 属性的值为 flex 或 inline-flex（行内 flex），表示容器为 flex 容器。flex 容器的所有子元素自动成为容器成员，即 flex 子元素。flex 容器具有 6 个属性，分别为 flex-direction、flex-wrap、flex-flow、justify-content、align-content、align-items。其中 flex-flow 为 flex-direction、flex-wrap 的简写形式。flex 子元素具有 6 个属性，用来设置项目的尺寸、位置、对齐方式等，分别为 order、flex-shrink、flex-grow、flex-basis、flex、align-self。其中 flex 为 flex-shrink、flex-grow、flex-basis 的简写形式。

【实例 2-9】制作 flex 布局页的导航。

修改 flex.vue 页面，在该页面中添加几个 navigator 组件，每个组件都用于导航到相应页面。对应页面演示容器属性或子元素属性的不同设置。演示效果如图 2-16 所示。

实例 2-9　　　　图 2-16　flex 布局页演示效果

具体代码（flex.vue 完整的代码）如下。

```html
<template>
    <view class="content">
        <view class="btnbox">
            <text class="title">容器属性</text>
            <navigator open-type="navigate" url="flexdirection">
                <button type="default" class="btn">flex-direction 属性</button>
            </navigator>
            <navigator open-type="navigate" url="flexwrap">
                <button type="default" class="btn">flex-wrap 属性</button>
            </navigator>
            <navigator open-type="navigate" url="justifycontent">
                <button type="default" class="btn">justify-content 属性</button>
            </navigator>
            <navigator open-type="navigate" url="alignitems">
                <button type="default" class="btn">align-items 属性</button>
            </navigator>
            <navigator open-type="navigate" url="aligncontent">
                <button type="default" class="btn">align-content 属性</button>
            </navigator>
        </view>

        <view class="btnbox">
            <text class="title">子元素属性</text>
            <navigator open-type="navigate" url="order" >
                <button type="default" class="btn" >order 属性</button>
            </navigator>
            <navigator open-type="navigate" url="flexshrink">
                <button type="default" class="btn">flex-shrink 属性</button>
            </navigator>
            <navigator open-type="navigate" url="flexgrow">
                <button type="default" class="btn">flex-grow 属性</button>
            </navigator>
            <navigator open-type="navigate" url="flexbasis">
                <button type="default" class="btn">flex-basis 属性</button>
            </navigator>
            <navigator open-type="navigate" url="alignself">
                <button type="default" class="btn">align-self 属性</button>
            </navigator>
        </view>
    </view>
</template>
<script>
export default {
    data() {
        return {
        }
    },
    methods:{
    }
}
</script>
<style>
```

```
.content {
    display:flex;
    flex-direction:column;
    align-items:center;
    justify-content:center;
    padding:30rpx;
}
.btnbox{
    display:flex;
    flex-direction:column;
    width:100%;
    margin-bottom:30rpx;
    padding-left:20rpx;
    padding-right:20rpx;
}
.title{
    font-size:20px;
    margin-bottom:20rpx;
}
.btn{
    margin-bottom:20rpx;
}
</style>
```

2.6.2 容器属性

1. flex-direction 属性

flex-direction 用于调整主轴的方向，可以调整为横向或者纵向。在默认情况下是横向，此时横轴为主轴，纵轴为交叉轴；如果改为纵向，则纵轴为主轴，横轴为交叉轴。其取值如表 2-13 所示。

表 2-13 flex-direction 属性值

值	描述
row	flex 子元素按横轴方向顺序排列（默认值）
row-reverse	flex 子元素按横轴方向逆序排列
column	flex 子元素按纵轴方向顺序排列
column-reverse	flex 子元素按纵轴方向逆序排列

【实例 2-10】演示 flex-direction 属性的不同取值的效果。

在 pages/flex 文件夹下新建 flexdirection.vue，演示 4 种方向的主轴的效果。演示效果如图 2-17 所示。

实例 2-10

注意：

新建页面后，需要在 pages.json 文件的 pages 节点中增加以下代码。

```
{
    "path" :"pages/flex/flexdirection",
    "style" :
```

```
{
    "navigationBarTitleText":"flex 布局",
    "enablePullDownRefresh":false
}
```

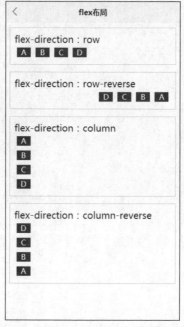

图 2-17　flexdirection.vue 页面演示效果

具体代码如下。

```
<template>
<view class="content">
    <view class="flex-direction">
        <text class="title">flex-direction:row</text>
        <view class="flex-row">
            <view class="item">A</view>
            <view class="item">B</view>
            <view class="item">C</view>
            <view class="item">D</view>
        </view>
    </view>
    <view class="flex-direction">
        <text class="title">flex-direction:row-reverse</text>
        <view class="flex-row-reverse">
            ……省略 4 个 view 组件中的代码
        </view>
    </view>
    <view class="flex-direction">
        <text class="title">flex-direction:column</text>
        <view class="flex-column">
            ……省略 4 个 view 组件中的代码
        </view>
```

```html
            </view>
            <view class="flex-direction">
                <text class="title">flex-direction:column-reverse</text>
                <view class="flex-column-reverse ">
                    ……省略4个view组件中的代码
                </view>
            </view>
        </view>
</view>
</template>
<script>
export default {
    data() {
        return {
        }
    },
    methods:{
        }
}
</script>
<style>
.content {
    display:flex;
    flex-direction:column;
    align-items:center;
    margin:20rpx auto;
    width:90%;
}
.flex-direction{
    display:flex;
    flex-direction:column;
    width:100%;
    margin-bottom:30rpx;
    border:1px solid #C7C7C7;
    padding:20rpx;
}
.flex-row{
    display:flex;
    width:100%;
}
.title{
    font-size:20px;
}
.item{
    width:30px;
    height:20px;
    margin:10rpx;
    text-align:center;
    background-color:#222222;
    line-height:20px;
    justify-content:space-between;
    color:#FFFFFF;
}
```

```
.flex-column{
    display:flex;
    flex-direction:column;
}
.flex-row-reverse{
    display:flex;
    flex-direction:row-reverse;
}
.flex-column-reverse{
    display:flex;
    flex-direction:column-reverse;
}
</style>
```

2. flex-wrap 属性

flex-wrap 属性用于在必要的时候将 flex 子元素换行，其取值如表 2-14 所示。

表 2-14　flex-wrap 属性值

值	描述
nowrap	容器为单行，在该情况下 flex 子元素可能会溢出容器。该值是默认属性值，其作用是不换行
wrap	容器为多行，flex 子元素溢出的部分会被放置到新行（换行），第一行显示在容器上方
wrap-reverse	反转 wrap 排列（换行），第一行显示在容器下方

【实例 2-11】使用 flex-wrap 属性设置矩形的换行和排序，效果如图 2-18 所示。

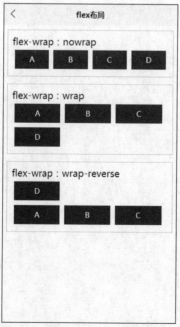

实例 2-11

图 2-18　flexwrap.vue 页面演示效果

具体代码如下。

```
<template>
<view class="content">
        <view class="flex-direction">
            <text class="title">flex-wrap:nowrap</text>
            <view class="flex-row">
                ……省略 4 个 view 组件中的代码,与实例 2-10 的一致
            </view>
        </view>
        <view class="flex-direction">
            <text class="title">flex-wrap:wrap</text>
            <view class="flex-row-wrap">
                ……省略 4 个 view 组件中的代码,与实例 2-10 的一致
            </view>
        </view>
        <view class="flex-direction">
            <text class="title">flex-wrap:wrap-reverse</text>
            <view class="flex-row-wrap-reverse">
                ……省略 4 个 view 组件中的代码,与实例 2-10 的一致
            </view>
        </view>
</view>
</template>
<script>                </script>
<style>
.content{
    display:flex;
    flex-direction:column;
    align-items:center;
    margin:20rpx auto;
    width:90%;
}
.flex-direction{
    display:flex;
    flex-direction:column;
    width:100%;
    margin-bottom:30rpx;
    border:1px solid #C7C7C7;
    padding:20rpx;
}
.flex-row{
    display:flex;
    width:100%;
}
.flex-row-wrap{
    display:flex;
    width:100%;
    flex-wrap:wrap;
}
.flex-row-wrap-reverse{
    display:flex;
    width:100%;
```

```
        flex-wrap:wrap-reverse;
}
.title{
        font-size:20px;
}
.item{
        width:100px;
        height:40px;
        margin:10rpx;
        text-align:center;
        background-color:#222222;
        line-height:40px;
        justify-content:space-between;
        color:#FFFFFF;
}
.flex-column{
        display:flex;
        flex-direction:column;
}
.flex-row-reverse{
        display:flex;
        flex-direction:row-reverse;
}
.flex-column-reverse{
        display:flex;
        flex-direction:column-reverse;
}
</style>
```

3. justify-content 属性

justify-content 属性用于设置子元素在主轴方向的排列方式，其取值如表 2-15 所示。

表 2-15　justify-content 属性值

值	描述
flex-start	flex 子元素将向主轴的起始位置对齐（默认值）
flex-end	flex 子元素将向主轴的结束位置对齐
center	flex 子元素将向主轴的中间位置对齐
space-between	flex 子元素沿主轴方向平均分布，第一个子元素的边界与主轴的起始位置边界对齐，最后一个子元素的边界与主轴的结束位置边界对齐
space-around	flex 子元素沿主轴方向平均分布，两端保留子元素与子元素间距大小的一半
space-evenly	子元素间距、第一个子元素距离主轴的起始位置的距离、最后一个子元素距离主轴的结束位置的距离均相等

【实例 2-12】演示 justify-content 不同属性值的效果。

在本例中，justifycontent.vue 页面中展示了主轴为横向时，justify-content 取不同值时子元素的排列方式。效果如图 2-19 所示。

实例 2-12

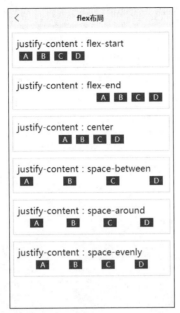

图 2-19　justifycontent.vue 页面演示效果

具体代码如下。

```
<template>
<view class="content">
    <view class="flex-direction">
        <text class="title">justify-content:flex-start</text>
        <view class="flex-row">
            <view class="item">A</view>
            <view class="item">B</view>
            <view class="item">C</view>
            <view class="item">D</view>
        </view>
    </view>
    <view class="flex-direction">
        <text class="title">justify-content:flex-end</text>
        <view class="flex-row flexend">
            ……省略 4 个 view 组件的代码，与实例 2-10 的一致
        </view>
    </view>
    <view class="flex-direction">
        <text class="title">justify-content:center</text>
        <view class="flex-row flexcenter">
            ……省略 4 个 view 组件的代码，与实例 2-10 的一致
        </view>
    </view>
    <view class="flex-direction">
        <text class="title">justify-content:space-between</text>
        <view class="flex-row flexbetween">
            ……省略 4 个 view 组件的代码，与实例 2-10 的一致
        </view>
    </view>
```

```
                <view class="flex-direction">
                    <text class="title">justify-content:space-around</text>
                    <view class="flex-row flexaround">
                        ……省略 4 个 view 组件的代码,与实例 2-10 的一致
                    </view>
                </view>
                <view class="flex-direction">
                    <text class="title">justify-content:space-evenly</text>
                    <view class="flex-row flexevenly">
                        ……省略 4 个 view 组件的代码,与实例 2-10 的一致
                    </view>
                </view>
        </view>
    </template>
    <script>           </script>
    <style>
……省略.content、.flex-direction、.flex-row、.title、.item 类的定义,它们与实例 2-10 的一致
        .flexend{
            justify-content:flex-end;
        }
        .flexcenter{
            justify-content:center;
        }
        .flexbetween{
            justify-content:space-between;
        }
        .flexaround{
            justify-content:space-around;
        }
        .flexevenly{
            justify-content:space-evenly;
        }
    </style>
```

4. flex-flow 属性

flex-flow 属性为 flex-direction 和 flex-wrap 属性的综合形式,默认值为 row nowrap,其语法格式如下。

```
flex-flow:column wrap;
```

5. align-items 属性

align-items 属性用于定义子元素在交叉轴上的对齐方式,其取值如表 2-16 所示。

表 2-16 align-items 属性值

值	描述
flex-start	flex 子元素向交叉轴的起始位置对齐
flex-end	flex 子元素向交叉轴的结束位置对齐
center	flex 子元素向交叉轴的中间位置对齐
baseline	子元素根据其基线进行对齐,在未单独设置基线时,等效于 flex-start
stretch	默认值。如果子元素未设置高度或设为 auto,将占满整个容器的高度

【实例 2-13】创建 alignitems.vue 页面，添加 A、B、C、D 这 4 个 view 组件，没有为它们设置高度，而设置外部 view 组件的高度为 50px，并添加蓝色边框。A1 view 组件高度为 30px，B1 view 组件高度为 50px，C1 view 组件高度为 40px，D1 view 组件高度为 60px。外层容器主轴为横向，交叉轴为纵向。演示效果如图 2-20 所示。

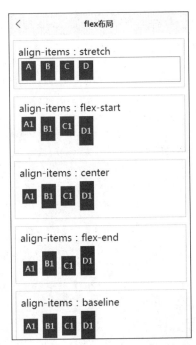

实例 2-13

图 2-20　alignitems.vue 页面演示效果

具体代码如下。

```
<template>
    <view class="content">
        <view class="flex-direction">
            <text class="title">align-items:stretch</text>
            <view class="flex-row stretch" style="height:50px; border:1px solid #007AFF;">
                <view class="item ">A</view>
                <view class="item ">B</view>
                <view class="item ">C</view>
                <view class="item ">D</view>
            </view>
        </view>
        <view class="flex-direction">
            <text class="title">align-items:flex-start</text>
            <view class="flex-row flexstart">
                <view class="item a">A1</view>
                <view class="item b">B1</view>
                <view class="item c">C1</view>
                <view class="item d">D1</view>
            </view>
```

```
                </view>
                <view class="flex-direction">
                    <text class="title">align-items:center</text>
                    <view class="flex-row flexcenter">
                        ……省略 A1、B1、C1、D1 这 4 个 view 组件的代码,与上一组相同
                    </view>
                </view>
                <view class="flex-direction">
                    <text class="title">align-items:flex-end</text>
                    <view class="flex-row flexend">
                        ……省略 A1、B1、C1、D1 这 4 个 view 组件的代码,与上一组相同
                    </view>
                </view>
                <view class="flex-direction">
                    <text class="title">align-items:baseline</text>
                    <view class="flex-row baseline">
                        ……省略 A1、B1、C1、D1 这 4 个 view 组件的代码,与上一组相同
                    </view>
                </view>
        </view>
</template>
<script>   </script>
<style>
.content {
        display:flex;
        flex-direction:column;
        align-items:center;
        margin:20rpx auto;
        width:90%;
    }
    .flex-direction{
        display:flex;
        flex-direction:column;
        width:100%;
        margin-bottom:30rpx;
        border:1px solid #C7C7C7;
        padding:20rpx;
    }
    .flex-row{
        display:flex;
        width:100%;
    }
    .title{
        font-size:20px;
    }
    .item{
        width:30px;
        /* height:20px; */
        margin:10rpx;
        text-align:center;
        background-color:#222222;
        /* line-height:20px; */
```

```
            justify-content:space-between;
            color:#FFFFFF;
    }
    .flexend{
            align-items:flex-end;
    }
    .flexcenter{
            align-items:center;
    }
    .flexstart{
            align-items:flex-start;
    }
    .stretch{
            align-items:stretch;
    }
    .baseline{
            align-items:baseline;
       }
    .a{
            height:30px;
            line-height:30px;
    }
    .b{
            height:50px;
            line-height:50px;
    }
    .c{
            height:40px;
            line-height:40px;
    }
    .d{
            height:60px;
            line-height:60px;
    }
</style>
```

6. align-content 属性

align-content 属性主要用于在进行多行排列时，设置子元素在交叉轴方向上的对齐方式。其取值有 stretch、flex-start、flex-end、center、space-between、space-around。其中 flex-start、flex-end、center 的含义与表 2-15 中的相同，space-between、space-around 的含义与表 2-15 中的类似，只是这里是交叉轴方向。

注意：

在进行多行排列时，要将 flex-wrap 属性设置为 wrap，表示允许换行。

【**实例 2-14**】创建 aligncontent.vue 页面，演示 align-content 属性的使用方法，效果如图 2-21 所示。此处，A1、B1、C1、D1、E1 的 height 属性未设置。

实例 2-14

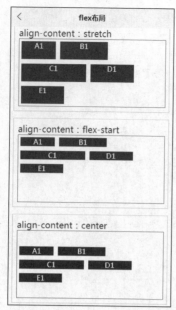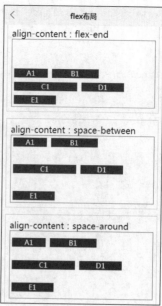

图 2-21 aligncontent.vue 页面演示效果

具体代码如下。

```
<template>
<view class="content">
    <view class="flex-direction">
      <text class="title">align-content:stretch</text>
      <view class="flex-row stretch"  >
        <view class="item a">A1</view>
        <view class="item b">B1</view>
        <view class="item c">C1</view>
        <view class="item d">D1</view>
        <view class="item e">E1</view>
      </view>
    </view>
    <view class="flex-direction">
      <text class="title">align-content:flex-start</text>
      <view class="flex-row flexstart">
        <view class="item a">A1</view>
        <view class="item b">B1</view>
        <view class="item c">C1</view>
        <view class="item d">D1</view>
        <view class="item e">E1</view>
      </view>
    </view>
    <view class="flex-direction">
      <text class="title">align-content:center</text>
      <view class="flex-row flexcenter">
        <view class="item a">A1</view>
        <view class="item b">B1</view>
        <view class="item c">C1</view>
        <view class="item d">D1</view>
        <view class="item e">E1</view>
```

```html
            </view>
        </view>
        <view class="flex-direction">
            <text class="title">align-content:flex-end</text>
            <view class="flex-row flexend">
                <view class="item a">A1</view>
                <view class="item b">B1</view>
                <view class="item c">C1</view>
                <view class="item d">D1</view>
                <view class="item e">E1</view>
            </view>
        </view>
        <view class="flex-direction">
            <text class="title">align-content:space-between</text>
            <view class="flex-row between">
                <view class="item a">A1</view>
                <view class="item b">B1</view>
                <view class="item c">C1</view>
                <view class="item d">D1</view>
                <view class="item e">E1</view>
            </view>
        </view>
        <view class="flex-direction">
            <text class="title">align-content:space-around</text>
            <view class="flex-row around">
                <view class="item a">A1</view>
                <view class="item b">B1</view>
                <view class="item c">C1</view>
                <view class="item d">D1</view>
                <view class="item e">E1</view>
            </view>
        </view>
    </view>
</template>
<script>  </script>
<style>
.content {
        display:flex;
        flex-direction:column;
        align-items:center;
        margin:20rpx auto;
        width:90%;

    }
    .flex-direction{
        display:flex;
        flex-direction:column;
        width:100%;
        margin-bottom:30rpx;
        border:1px solid #C7C7C7;
        padding:20rpx;
    }
    .flex-row{
        display:flex;
```

```css
            width:100%;
            flex-wrap:wrap;
            height:150px;
            border:1px solid #0066CC;
        }
        .title{
            font-size:20px;
        }
        .item{
            width:30px;
            height:20px;
            margin:10rpx;
            text-align:center;
            background-color:#222222;

            line-height:20px;
            justify-content:space-between;
            color:#FFFFFF;
        }
.stretch{
    align-content:stretch;
}
.flexstart{
    align-content:flex-start;
}
.flexend{
    align-content:flex-end;
}
.flexcenter{
    align-content:center;
}
.between{
    align-content:space-between;
}
.around{
    align-content:space-around;
}

    .a{
        width:80px;
    }
    .b{
        width:110px;
    }
    .c{
        width:150px;
    }
    .d{
        width:100px;
    }
    .e{
        width:100px;
    }
</style>
```

2.6.3 子元素属性

1. order 属性

order 属性用于定义子元素的排列顺序。属性值越小，排序越靠前，默认值为 0。

【实例 2-15】创建 order.vue 页面，在 order.vue 页面中放置 4 个 view 组件，并设置它们的 order 属性值为 6、3、2、5，效果如图 2-22 所示。

具体代码如下。

图 2-22 order.vue 页面演示效果

<template></template>模块：

```
<template>
<view class="content">
    <view class="flex-direction">
        <text class="title">order</text>
        <view class="flex-row ">
            <view class="item order-6">A6</view>
            <view class="item order-3">B3</view>
            <view class="item order-2">C2</view>
            <view class="item order-5">D5</view>
        </view>
    </view>
</view>
</template>
```

<style></style>模块：

```
<style>
    ……这里的.content、.flex-direction、.flex-row、.title、.item 类与实例 2-10 的一致，故省略
    .order-6{
        order:6;
    }
    .order-2{
        order:2;
    }
    .order-3{
        order:3;
    }
    .order-5{
        order:5;
    }
</style>
```

2. flex-shrink 属性

flex-shrink 属性用于定义子元素的收缩比例。当子元素在主轴方向上溢出时，根据子元素 flex-shrink 属性指定的比例收缩子元素以适应容器。flex-shrink 属性值为非负数，默认为 1，若为 0 表示不收缩。

【实例 2-16】创建 flexshrink.vue 页面，在 flexshrink.vue 页面中放置 3 个 view 组件，然

后设置其 flex-shrink 属性，查看收缩效果，如图 2-23 所示。其中 A、B、C 这 3 个 view 组件的宽度为 100px，外层容器的宽度为 200px。第一组 view 组件的 flex-shrink 都为默认值 1，所以它们的收缩比例一样大；第二组 view 组件的 flex-shrink 为 0，表示不收缩，所以 C 在容器外面；第三组 view 组件 A、B、C 的 flex-shrink 的值分别为 1、2、3，将溢出宽度 100px 分成 6 份，A 收缩 1 份，B 收缩 2 份，C 收缩 3 份。

具体代码如下。

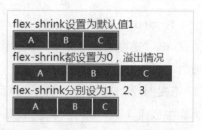

图 2-23 flexshrink.vue 页面演示效果

<template></template>模块：

```html
<template>
<view class="content">
    <view class="flex-direction">
        <text class="title">flex-shrink 设置为默认值 1</text>
        <view class="flex-row1 ">
            <view class="item1 ">A</view>
            <view class="item1 ">B</view>
            <view class="item1 ">C</view>
        </view>
        <text class="title">flex-shrink 都设置为 0，溢出情况</text>
        <view class="flex-row1 ">
            <view class="item1 flexshrink-0">A</view>
            <view class="item1 flexshrink-0">B</view>
            <view class="item1 flexshrink-0">C</view>
        </view>
        <text class="title">flex-shrink 分别设为 1、2、3</text>
        <view class="flex-row1 ">
            <view class="item1 flex-shrink-1">A</view>
            <view class="item1 flex-shrink-2">B</view>
            <view class="item1 flex-shrink-3">C</view>
        </view>
    </view>
</view>
</template>
```

<style></style>模块：

```
<style>
……这里的.content、.flex-direction、.title 类定义与实例 2-10 的一致，故省略
    .flex-row1{
        display:flex;
        width:200px;
        border:2px solid #007AFF;
    }
    .item1{
        width:100px;
        height:30px;
        line-height:30px;
        text-align:center;
```

```
            background-color:#222222;
            line-height:30px;
            color:#FFFFFF;
            border:1px solid #FFFFFF;
        }
        .flexshrink-0{
            flex-shrink:0;
        }
        .flex-shrink-1{
            flex-shrink:1;
        }
        .flex-shrink-2{
            flex-shrink:2;
        }
        .flex-shrink-3{
            flex-shrink:3;
        }
</style>
```

3. flex-grow 属性

flex-grow 属性用于定义子元素的扩张比例，当主轴方向上还有剩余空间时，根据子元素的 flex-grow 属性指定的比例扩张子元素来分配剩余空间。flex-grow 属性值为非负数，默认为 0，表示不扩张。

【实例 2-17】创建 flexgrow.vue 页面，在 flexgrow.vue 页面中，设置外层容器宽度为 300px，内部组件 A、B、C 宽度为 50px。在第一组 view 组件中，flex-grow 的值为 0，不扩张。在第二组 view 组件中，flex-grow 的值分别为 0、1、2，所以 A 不扩张，B 的扩张大小为剩余空间 150px 的 1/3，即 50px（B 扩张后的实际宽度为 100px），C 的扩张大小为剩余空间 150px 的 2/3，即 100px（C 扩张后的实际宽度为 150px）。具体演示效果如图 2-24 所示。

具体代码如下。

<template></template>模块：

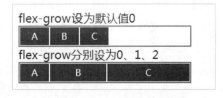

图 2-24　flexgrow.vue 页面演示效果

```
<template>
<view class="content">
    <view class="flex-direction">
        <text class="title">flex-grow设为默认值0</text>
        <view class="flex-row2 ">
            <view class="item1 width-50">A</view>
            <view class="item1 width-50">B</view>
            <view class="item1 width-50">C</view>
        </view>
        <text class="title">flex-grow分别设为0、1、2</text>
        <view class="flex-row2 ">
            <view class="item1 width-50 grow-0">A</view>
            <view class="item1 width-50 grow-1">B</view>
            <view class="item1 width-50 grow-2">C</view>
        </view>
```

```
            </view>
        </view>
</template>
```

<style></style>模块：

```
<style>
……这里的.content、.flex-direction、.title、.item1类定义与实例2-16的一致，故省略
        .flex-row2{
                display:flex;
                width:300px;
                border:2px solid #007AFF;
        }
        .width-50{
                width:50px;
        }
        .grow-0{
                flex-grow:0;
        }
        .grow-1{
                flex-grow:1;
        }
        .grow-2{
                flex-grow:2;
        }
</style>
```

4. flex-basis 属性

flex-basis 属性主要用于根据主轴方向替换子元素的宽度或高度属性。当容器 flex-direction 属性值为 row 或 row-reverse 时，如果子元素的 flex-basis 属性和 width 属性同时指定了数值，则使用 flex-basis 属性代替 width 属性。当容器 flex-direction 属性值为 column 或 column-reverse 时，如果子元素的 flex-basis 属性和 height 属性同时指定了数值，则使用 flex-basis 属性代替 heihgt 属性。flex-basis 属性的语法格式如下。

```
flex-basis:auto(默认值) | <数字>px;
```

数字比 auto 的优先级高，如果 flex-basis 属性值为 auto，而 width 或 height 属性值是数字，则采用数字作为最终属性值。

【实例 2-18】创建 flexbasis.vue 页面，在 flexbasis.vue 页面中，放置 A、B、C 这 3 个 view 组件，其宽度都设置为 50px，其中 A 的 flex-basis 属性值为 150px，B、C 的 flex-basis 属性值为默认值 auto。演示效果如图 2-25 所示。

图 2-25　flexbasis.vue 页面的演示效果

具体代码如下。

<template></template>模块：

```
<template>
<view class="content">
        <view class="flex-direction">
                <text class="title">A 的 flex-basis 属性值为 150px</text>
                <view class="flex-row2 ">
```

```
                <view class="item2  flexbasis-150">A</view>
                <view class="item2 ">B</view>
                <view class="item2 ">C</view>
            </view>
        </view>
    </view>
</template>
```

<style></style>模块：

```
<style>
    ……这里的.content、.flex-direction、.title类定义与实例2-16的一致,故省略。.flex-row2
与实例2-17的一致, .item2与.item1的差别为width, 这里width为50px

    .flexbasis-150{
        flex-basis:150px;
    }
</style>
```

5. flex 属性

flex 属性是 flex-grow、flex-shrink 和 flex-basis 属性的综合简写，默认值为 0 1 auto，其中后两个值是可选的。该属性有两个快捷值——auto 和 none，其中 auto 代表 1 1 auto，none 代表 0 0 auto。

建议优先使用 flex 属性，而不是单独使用 3 个分离的属性。

6. align-self 属性

align-self 属性用来设置子元素在交叉轴方向上的对齐方式，从而覆盖容器的 align-items 属性。其取值与 align-items 的相同，如表 2-16 所示。其默认值为 auto。

【实例 2-19】创建 alignself.vue 页面，在 alignself.vue 页面中，设置 A、B、C、D 这 4 个组件的 align-self 属性值分别为 flex-start、flex-end、center、stretch。A、B、C 的高度都在.item1 中定义为 30px，D 的高度通过 style="height: auto;"设置为 auto，其优先级高于.item1 的高度值 30px。演示效果如图 2-26 所示。

图 2-26　alignself.vue 页面演示效果

具体代码如下。

<template></template>模块：

```
<template>
<view class="content">
    <view class="flex-direction">
        <text class="title">align-self</text>
        <view class="flex-row2 height-250 ">
            <view class="item1 align-self-start">A</view>
            <view class="item1 align-self-end">B</view>
            <view class="item1 align-self-center">C</view>
            <view class="item1 align-self-stretch" style="height:auto;" >D</view>
        </view>
    </view>
</view>
</template>
```

<style></style>模块：

```
<style>
……这里的.content、.flex-direction、.title、.item1 定义与实例2-16的一致，故省略
    .height-250{
        height:80px;
    }
    .align-self-center{
        align-self:center;
    }
    .align-self-start{
        align-self:flex-start;
    }
    .align-self-end{
        align-self:flex-end;
    }
    .align-self-stretch{
        align-self:stretch;
    }
</style>
```

2.7 案例：页面导航

本案例主要利用flex布局设计一个类似于支付宝、美团、完美校园等App的常用的页面导航。页面效果如图2-27所示。

案例：页面导航

图2-27 页面效果

实现步骤

（1）新建项目。在pages/index目录下新建一个CSS文件index.css，在index.vue中的<style></style>部分添加导入语句。修改pages.json文件中index.vue页面的navigationBarTitleText的值为"XX应用"。

```
<style>
    @import url("index.css");
</style>
```

（2）设置页面的布局。

```
<template>
        <view class="content">
        </view>
</template>
```

在 index.css 文件中添加以下代码。

```
.content {
    display:flex;
    flex-direction:column;
    align-items:center;
    justify-content:space-between;
}
```

（3）准备资源。在 iconfont 网站上下载图标，并将其复制到 static/icon 目录中。

（4）制作搜索框。参考 3.3.1 小节安装 uni-ui 中的搜索组件。

<template></template>视图层部分的代码如下。

```
<view class="search-box">
    <view class="search-bar">
        <uni-search-bar placeholder="热门电影：独行月球" @confirm="search" clearButton="none" cancelButton="none">
        </uni-search-bar>
        <view class="search-boxht">
            <image src="@/static/icon/huatong.png"></image>
        </view>
    </view>
    <view class="search-icon">
        <image src="@/static/icon/my.png"></image>
        <image src="@/static/icon/jia.png"></image>
    </view>
</view>
```

CSS 代码如下。

```
.search-box {
    width:100%;
    background-color:#1977ff;
    border:#1977ff solid;
    display:flex;
    flex-direction:row;
    align-items:center;
    justify-content:space-between;
}
.search-bar {
    flex-grow:1;
    position:relative;
}
.search-bar .search-boxht {
    position:absolute;
    top:30%;
    left:83%;
}
.search-icon {
```

```
        display:flex;
        flex-direction:row;
        align-items:center;
        justify-content:space-around;
        padding-right:30rpx;
}
.search-box image {
        z-index:1;
        width:25px;
        height:25px;
        margin-top:2px;
        margin-left:10px;
}
```

(5)制作"扫一扫""付钱""收钱""卡包"部分,代码如下。

```
<view class="black-box">
    <view class="black-icon">
        <view class="icon-flex">
            <image src="@/static/icon/saoyisao.png"></image>
            <text>扫一扫</text>
        </view>
        <view class="icon-flex">
            <image src="@/static/icon/money.png"></image>
            <text>付钱</text>
        </view>
        <view class="icon-flex">
            <image src="@/static/icon/money.png"></image>
            <text>收钱</text>
        </view>
        <view class="icon-flex">
            <image src="@/static/icon/kabao.png"></image>
            <text>卡包</text>
        </view>
    </view>
</view>
```

CSS 样式代码如下。

```
.black-box {
        width:100%;
        background-color:#1977ff;
        border:#1977ff solid;
}
.black-icon {
        margin-bottom:15px;
        display:flex;
        flex-direction:row;
        justify-content:space-around;
}
.icon-flex {
        display:flex;
        flex-direction:column;
        align-items:center;
        justify-content:center;
}
```

```
.icon-flex image {
    z-index:1;
    width:30px;
    height:30px;
    margin:0 auto;
}
.icon-flex text {
    font-size:12px;
    color:white;

}
```

(6)导航页面第一行的代码如下,其他行的代码类似。

<template></template>视图层部分的代码如下。

```
    <view class="white-box">
        <view class="white-flex1">
            <view class="icon-flex1">
                <image src="@/static/icon/koubei.png"></image>
                <text>口碑</text>
            </view>
            <view class="icon-flex1">
                <image src="@/static/icon/CivicCenter.png"></image>
                <text>市民中心</text>
            </view>
            <view class="icon-flex1">
                <image src="@/static/icon/zhuanzhang2.png"></image>
                <text>转账</text>
            </view>
            <view class="icon-flex1">
                <image src="@/static/icon/creditcard.png"></image>
                <text>还款</text>
            </view>
            <view class="icon-flex1">
                <image src="@/static/icon/elma.png"></image>
                <text>饿了么</text>
            </view>
        </view>
</view>
```

CSS样式代码如下。

```
.white-box {
    width:99%;
    height:490rpx;
    background-color:#fafafa;
}
.icon-flex1 {
    width:50px;
    text-align:center;
}
.icon-flex1 image {
    display:flex;
    flex-direction:column;
    z-index:1;
```

```
        width:25px;
        height:25px;
        margin:0 auto;
}
.icon-flex1 text {
        font-size:10px;
}
.white-flex1 {
        display:flex;
        flex-direction:row;
        justify-content:space-around;
        margin-top:15px;
}
```

本章小结

本章主要介绍了 uni-app 的 pages.json 文件、资源引用、页面样式、尺寸单位、常用的 4 个基础组件、属性绑定和事件绑定、v-for 渲染数据，以及移动应用开发中常用的 flex 布局，最后用一个类似支付宝首页的案例综合应用了相关知识点。

项目实战

实现一个模拟影院购票的系统，页面效果如图 2-28 所示。该项目中会用到 JS 代码，通过 uni.showToast()弹出提示框。

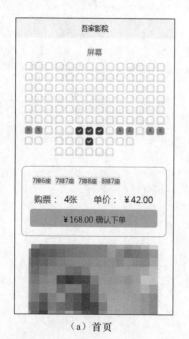

（a）首页　　　　　　　　　　　（b）下单提示

图 2-28　页面效果

拓展实训项目

党的二十大报告提出"完善志愿服务制度和工作体系。弘扬诚信文化，健全诚信建设长效机制"。现在，越来越多的大学生的志愿服务意识不断增强，积极参与到志愿者服务中。请设计一个志愿者服务小程序的"志愿者活动页"，具体包括搜索框、活动列表，活动列表信息包括活动标题、承办单位、活动开始时间、人员要求和【报名】按钮。点击列表项，跳转至活动详情页，点击列表项中的【报名】按钮，提示"报名成功"信息。在搜索框中输入搜索内容后，点击软键盘中的【搜索】按钮，刷新志愿者活动页。

第3章
uni-app基础扩展

本章导读

　　本章主要讲解 uni-app 的生命周期、条件编译、扩展组件 uni-ui 的安装和使用等内容。通过对本章的学习，读者能够初步掌握移动应用页面的设计。

学习目标

知识目标	1. 掌握 uni-app 的生命周期 2. 掌握条件编译 3. 掌握 uni-ui 等扩展组件的安装和使用
能力目标	1. 能够熟练使用扩展组件搭建页面 2. 能够实现跨平台处理 3. 能够理解生命周期函数的应用场景
素质目标	1. 具有良好的软件编码规范素养 2. 具有探索新知、不畏困难的精神 3. 具有强烈的爱国情感和民族自豪感

知识思维导图

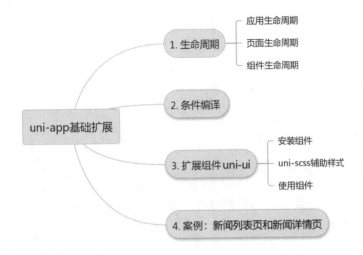

3.1 生命周期

uni-app 生命周期分为应用生命周期、页面生命周期和组件生命周期,用得较多的是页面生命周期。

3.1.1 应用生命周期

uni-app 支持的应用生命周期函数如表 3-1 所示。

表 3-1 应用生命周期函数

函数名	说明
onLaunch	当 uni-app 初始化完成时触发(全局只触发一次)
onShow	当 uni-app 启动,或从后台进入前台显示时触发
onHide	当 uni-app 从前台进入后台时触发
onError	当 uni-app 报错时触发
onUniNViewMessage	对 nvue 页面发送的数据进行监听
onUnhandledRejection	对未处理的 Promise 拒绝事件监听函数
onPageNotFound	页面不存在监听函数
onThemeChange	监听系统主题变化

应用生命周期函数只能在 App.vue 中监听,在其他页面中无效。App.vue 是 uni-app 的主组件,所有页面都是在 App.vue 中进行切换的,它是页面入口文件。App.vue 默认有 3 个生命周期函数 onLaunch、onShow、onHide,代码见实例 3-1,这 3 个函数在开发中经常使用。

【实例 3-1】使用默认模板新建一个 uni-app 项目 uniappch03，本章实例将使用本项目。查看 App.vue 的代码。

实例 3-1

```
<script>
    export default {
        onLaunch:function() {
            console.log('App Launch')
        },
        onShow:function() {
            console.log('App Show')
        },
        onHide:function() {
            console.log('App Hide')
        }
    }
</script>
```

选择以真机方式运行程序，打开控制台，会发现 onLaunch 函数只触发一次，onShow 函数和 onHide 函数只要切换前后台就会触发。

3.1.2 页面生命周期

uni-app 提供了大量的页面生命周期函数，写在 pages 文件夹的页面文件中。以下列举了项目中经常用到的页面生命周期函数，如表 3-2 所示。

表 3-2 页面生命周期函数

函数名	说明
onLoad	监听页面加载，其参数为上一个页面传递的参数，参数类型为 Object
onShow	监听页面显示。页面每次出现在屏幕上都会触发，包括从下级页面点返回露出当前页面
onReady	监听页面初次渲染完成。注意，如果渲染速度快，会在页面进入动画完成前触发
onHide	监听页面隐藏
onUnload	监听页面卸载
onPullDownRefresh	监听用户下拉动作，一般用于下拉刷新
onReachBottom	监听页面滚动到底部（不是 scroll-view 滚到底）的事件，常用于下拉显示下一页数据

1. onLoad 函数

onLoad 函数用于监听页面加载，其参数是上一个页面传递的参数。onLoad 函数比较重要，它一般有两个作用。

➢ 调用数据接口，获取服务器端数据。
➢ 接收上一个页面传递的参数。

【实例 3-2】演示接收上一个页面传递的参数。

在 uniappch03 项目中，在 index.vue 页面中添加一个按钮，实现点击该按钮跳转到 testload.vue 页面。

实例 3-2

index.vue 中添加的代码涉及以下两个模块。

（1）<template></template>模块。

```
<button type="primary" size="mini" @click="nextpage">跳到下一个页面</button>
```

（2）<script></script>模块，实现跳转到 testload.vue 页面，并传递两个参数。

```
methods:{
    nextpage(){
        uni.navigateTo({
            url:'testload?id=1&name=uniapp'
        })
    }
}
```

（3）在 index 目录下新建 testload.vue 页面，在<script></script>添加变量定义和 onLoad 函数。具体代码如下。

```
<script>
    export default {
        data() {
            return {
                id: '',
                name:''
            }
        },
        onLoad: function (option) {
            this.id = option.id;
            this.name = option.name;
        }
    }
</script>
```

在<template></template>添加如下代码。

```
<text>这是 testload.vue 页面。{{id}}------{{name}} </text>
```

演示效果如图 3-1 所示。

（a）index.vue 页面　　　　　　（b）testload.vue 页面

图 3-1　演示效果（1）

2. onPullDownRefresh 函数

onPullDownRefresh 函数用于监听用户下拉动作，实现下拉刷新。

【实例 3-3】实现下拉刷新数据。

在 uniappch03 中的 index.vue 中添加 text 组件，显示 newsList 中的数据，下拉后，会更新 newList 的数据。演示效果如图 3-2 所示。

（a）下拉前

（b）下拉后

图 3-2　演示效果（2）

具体的代码如下。

```
<template>
    <view class="content">
        ……省略页面中其他内容
        <text v-for="(item, i) in newsList" :key="i">
            {{item}}
        </text>
    </view>
</template>
<script>
    export default {
        data() {
            return {
                title:'Hello',
                newsList:['1.感悟非凡成就，凝聚奋进力量','2.践行健康生活方式蔚然成风','3.世界感受中国非凡十年巨变']
            }
        },
        onLoad() {
        },
        onPullDownRefresh(){
            this.newsList = ['1.坚守平凡岗位，谱写生命华章','2.凝聚党心民心，永葆生机活力','3.稳健持重，开拓进取'];
            uni.stopPullDownRefresh();
        },
        methods:{
            ……省略实例 3-2 中的 nextpage 方法
        }
    }
</script>
……省略<style></style>模块的代码
```

注意：

（1）onPullDownRefresh 与 onLoad 平级。

（2）在 pages.json 中找到当前页面，在 style 属性中开启下拉刷新功能，其代码如下。

```json
{
            "path":"pages/index/index",
            "style":{
                "navigationBarTitleText":"uni-app",
                "enablePullDownRefresh":true

            }
        }
```

（3）数据刷新后，使用 uni.stopPullDownRefresh 方法关闭下拉刷新功能。

3. onReachBottom 函数

onReachBottom 是上拉加载生命周期函数，往往用来加载下一页的数据。

【实例 3-4】演示上拉加载生命周期函数 onReachBottom 的使用。

首先增加 newsList 中的数据，让页面中出现滚动条，然后在 onReachBottom 函数中添加数据。具体代码如下。

实例 3-3 和
实例 3-4

```
<script>
    export default {
        data() {
            return {
                title:'Hello',
                newsList:['1.感悟非凡成就，凝聚奋进力量','2.践行健康生活方式蔚然成风','3.世界感受中国非凡十年巨变','新闻标题','新闻标题','新闻标题','新闻标题','新闻标题','新闻标题','新闻标题'],
            }
        },
        onLoad() {

        },
        onPullDownRefresh(){
            this.newsList = ['1.坚守平凡岗位，谱写生命华章','2.凝聚党心民心，永葆生机活力','3.稳健持重，开拓进取'];
            uni.stopPullDownRefresh();
        },
        onReachBottom(){
            console.log("页面触底");
            this.newsList = [...this.newsList,...['3-新闻','4-新闻']]
        },
        methods:{
            nextpage(){
                uni.navigateTo({
                    url:'testload?id=1&name=uniapp'
                })
            }
        }
    }
</script>
```

演示效果如图 3-3 所示。后 4 行为上拉后增加的数据。

（a）上拉前

（b）上拉后

图 3-3 演示效果（3）

3.1.3 组件生命周期

uni-app 组件的生命周期与 Vue 标准组件的生命周期相同。组件生命周期函数写在 components 文件夹下的文件中。Vue2 的组件生命周期函数如表 3-3 所示。Vue3 的组件生命周期函数可参考 Vue.js 的官方网站。

表 3-3 Vue2 的组件生命周期函数

函数名	说明
beforeCreate	在实例初始化之前被调用
created	在实例创建完成后被立即调用
beforeMount	在挂载开始之前被调用
mounted	在挂载到实例上之后被调用
beforeUpdate	在数据更新时调用，发生在虚拟 DOM 打补丁之前
updated	在由于数据更改导致虚拟 DOM 重新渲染和打补丁之后调用
beforeDestroy	在实例销毁之前调用
destroyed	在 Vue 实例销毁后调用

3.2 条件编译

uni-app 能实现一套代码多平台运行，其核心是通过编译器和运行时来实现。

编译器：用于将 uni-app 统一代码编译生成每个平台支持的特有代码，如在小程序平台，编译器将 VUE 文件拆分成 WXML、WXSS、JS 等代码。

运行时：用于动态处理数据绑定、事件代理，保证 VUE 文件和平台宿主数据的一致性。

uni-app 已将常用的组件、JS API 封装到框架中，开发者按照 uni-app 规范进行开发即可保证多平台兼容性，可直接满足大部分业务需求。但每个平台有自己的一些特性，因此会存在一些无法跨平台的情况。这就需要用到条件编译。

条件编译是指用特殊的注释作为标记，在编译时根据这些特殊的注释，将注释里面的代码编译到不同平台。条件编译以 #ifdef 或 #ifndef 加 %PLATFORM% 开头，以 #endif 结尾。

> #ifdef: if defined 表示仅在某平台存在。
> #ifndef: if not defined 表示除了某平台均存在。
> %PLATFORM%表示平台名称，其值及生效条件如表 3-4 所示。

表 3-4 %PLATFORM%的值及生效条件

值	生效条件
VUE3	HBuilderX 3.2.0+
APP-PLUS	App
APP-PLUS-NVUE 或 APP-NVUE	App nvue 页面
H5	H5
MP-WEIXIN	微信小程序
MP-ALIPAY	支付宝小程序
MP-BAIDU	百度小程序
MP-TOUTIAO	字节跳动小程序
MP-LARK	飞书小程序
MP-QQ	QQ 小程序
MP-KUAISHOU	快手小程序
MP-JD	京东小程序
MP-360	360 小程序
MP	微信小程序/支付宝小程序/百度小程序/字节跳动小程序/飞书小程序/QQ 小程序/快手小程序/京东小程序/360 小程序
QUICKAPP-WEBVIEW	快应用通用（包含联盟、华为）
QUICKAPP-WEBVIEW-UNION	快应用联盟
QUICKAPP-WEBVIEW-HUAWEI	快应用华为

条件编译支持的文件有 VUE 文件、JS 文件、CSS 文件、pages.json 文件和各种预编译语言文件，如 SCSS 文件、LESS 文件、STYLUS 文件、TS 文件、PUG 文件。

条件编译是利用注释实现的，在不同语法里注释的写法不一样，JS 中使用"//注释"，CSS 中使用"/* 注释 */"，Vue/nvue 模板里使用"<!-- 注释 -->"。

1. 控制页面

在 VUE 文件的<template></template>标签内，可使用"<!-- 注释 -->"控制页面。

【实例 3-5】演示控制页面的方法。

在 uniappch03 项目的 index.vue 页面中的<template></template>标签内修改<image></image>标签相关内容，代码如下：

```
<template>
    <view class="content">
        <!-- #ifdef H5 -->
        <image class="logo" src="/static/htmlbig.png"></image>
```

```
            <!-- #endif -->
            <!-- #ifdef MP-WEIXIN -->
                <image class="logo" src="/static/weixin.jpeg"></image>
            <!-- #endif -->
            <view class="text-area">
                <text class="title">{{title}}</text>
            </view>
            <button type="primary" size="mini" @click="nextpage">跳到下一个页面</button>
            <text v-for="(item, i) in newsList" :key="i">
                {{item}}
            </text>
        </view>
</template>
```

H5 端的演示效果如图 3-4 所示,微信小程序端的演示效果如图 3-5 所示。

图 3-4 H5 端演示效果

图 3-5 微信小程序端演示效果

2. 控制 CSS 样式

在<style></style>标签内或 CSS 文件内,条件编译使用 "/* 注释 */"。

【实例 3-6】演示控制 CSS 样式的方法。

在本例中,在 H5 平台使用红色文字,在微信小程序平台使用黑色文字。

修改 index.vue 中<template></template>标签内的<text></text>标签,代码如下。

```
<text v-for="(item, i) in newsList" :key="i" class="newmain">
        {{item}}
</text>
```

在<style></style>标签内添加如下代码。

```
/* #ifdef H5 */
.newmain{
    color:red;
}
/* #endif */
/* #ifdef MP-WEIXIN */
.newmain{
    color:black;
}
/* #endif */
```

分别在 H5 和微信小程序端浏览页面,在 H5 端,页面的"新闻标题"部分显示为红色文字,在微信小程序端,页面的"新闻标题"部分显示为黑色文字。请读者自行查看。

3. 控制 JS

在<script></script>标签内或者 JS 文件内,条件编译使用"//注释"。

【实例 3-7】演示控制 JS 的方法。

实例 3-5、
实例 3-6 和
实例 3-7

在本例中,在 H5 平台,testload.vue 页面显示"1——Html 平台";在微信小程序端,testload.vue 页面显示"1——小程序平台"。具体演示效果请读者自行运行查看。

修改 index.vue 中的<script></script>标签内的 nextpage 方法,代码如下。

```
methods:{
    nextpage(){
        // #ifdef H5
        uni.navigateTo({
            url:'testload?id=1&name=Html 平台'
        })
        // #endif
        // #ifdef MP-WEIXIN
        uni.navigateTo({
            url:'testload?id=1&name=小程序平台'
        })
        // #endif
    }
}
```

以上实例都只用了#ifdef,表 3-5 列举了条件编译其他的语法结构。

表 3-5 条件编译其他的语法结构

语法结构	说明
#ifdef APP-PLUS 需条件编译的代码 #endif	用于仅存在于 App 平台的代码
#ifndef H5 需条件编译的代码 #endif	用于除了 H5 平台,其他平台均存在的代码
#ifdef H5 || MP-WEIXIN 需条件编译的代码 #endif	用于在 H5 平台或微信小程序平台存在的代码

3.3 扩展组件 uni-ui

uni-app 提供了一系列内置组件,在 2.5 节已介绍部分组件,更多组件将在第 4 章中详细介绍。本节讲解扩展组件的使用。uni-app 常用的扩展组件有 uni-ui、uView、iView 等。本节主要以 uni-ui 为例进行讲解,其他扩展组件的用法读者可以参考对应的官方文档。

uni-ui 是 DCloud 提供的一个跨平台 UI 库，它是基于 Vue 组件和 flex 布局的、无 DOM 的跨多平台 UI 框架。uni-ui 不包括基础组件，它是基础组件的补充。

3.3.1 安装组件

本节主要介绍安装组件的 3 种方式，分别是 uni_modules 插件方式、easycom 组件规范方式和 npm 方式。

1. uni_modules 插件方式

uni_modules 是 uni-app 的插件模块化规范（HBuilderX 3.1.0+支持），通常是对一组 JS SDK、组件、页面、uniCloud 云函数、公共模块等的封装，用于嵌入 uni-app 项目中使用，也支持直接封装为项目模板。

插件开发者可以像开发 uni-app 项目一样编写一个 uni_modules 插件，并在 HBuilderX 中将其直接上传至插件市场。

插件使用者可以在插件市场中查找符合自己需求的 uni_modules 插件，使用 HBuilderX 3.1.0+将其直接导入自己的 uni-app 项目中。后续还可以在 HBuilderX 中直接升级插件。

（1）uni-ui 组件完全安装

在 uni-app 官网的 uni-ui 组件中【使用组件】栏可以链接到 uni-ui 插件页面。也可以在 HBuilderX 中选择【工具】—【插件安装】—【安装新插件】—【前往插件市场安装】，【插件安装】对话框如图 3-6 所示。

图 3-6 【插件安装】对话框

打开插件市场页面，选择导航栏的【前端组件】，然后搜索"uni-ui"，打开 uni-ui 插件的页面，如图 3-7 所示。

图 3-7 uni-ui 插件的页面

单击【使用 HBuilderX 导入插件】按钮，弹出提示框，如图 3-8 所示。

单击【打开 URL:HBuilderX】按钮，弹出项目选择相关对话框，如图 3-9 所示。如果没有登录 HBuilderX，则需要登录。

图 3-8 提示框

图 3-9 项目选择相关对话框

选择项目后，HBuilderX 将会安装对应插件，并安装在项目根目录下的 uni-modules 文件夹中。

注意：

uni-ui 组件依赖 scss，必须安装 scss 插件，uView 才能正常运行。在插件市场找到 scss/sass 编译插件进行安装。

【实例 3-8】演示 uni-ui 的使用。

在 uniappch03 项目中采用 uni-modules 插件方式安装 uni-ui 组件，修改 index.vue 页面，为"Hello"部分添加一个徽章效果。具体代码如下。

<template></template>标签内的代码如下。

```
<view class="text-area">
    <uni-badge    class="uni-badge-left-margin"   text="2"   type="warning"
absolute="rightTop" size="normal">
```

```
                <view class="box">
                    <text >{{title}}</text>
                </view>
            </uni-badge>
</view>
```

在<style></style>标签内,添加样式类.box,代码如下。

```
.box {
    height:40px;
    display:flex;
    justify-content:center;
    align-items:center;
    text-align:center;
    background-color:#000000;
    color:#fff;
    font-size:36rpx;
    margin-bottom:10px;
    padding:0 20rpx ;
}
```

修改部分的效果如图3-10所示。

(2)单个组件安装

uni-ui组件可以不安装所有组件,而仅安装个别需要的组件。
uni-app通过uni_modules单独安装某个需要的组件的方式,和安装完整的uni-ui组件的方式一样,可以通过uni-app官网链接打开组件页面进行安装,也可以在"插件市场"的前端组件栏搜索对应的组件进行安装。

图3-10 修改部分的效果

通过uni_modules插件方式安装组件时,会自动安装其相关组件。组件安装后直接使用即可,无须导入和注册。

2. easycom 组件规范方式

传统Vue组件需要经过安装、引用、注册这3个步骤后才能使用,easycom将其精简为1步。

只要组件安装在项目的components目录下或uni_modules目录下,并符合"components/组件名称/组件名称.vue"或"uni_modules/组件名称/组件名称.vue"的目录结构,就可以不用引用、注册,直接在页面中使用。

在图3-7所示的插件页面中单击【下载插件ZIP】按钮,直接下载插件,并将其按照easycom组件规范放入components或uni_modules目录中即可使用。

easycom是自动开启的,有需求时可以在pages.json的easycom节点进行个性化配置。easycom属性如表3-6所示。

表3-6 easycom属性

属性	类型	默认值	描述
autoscan	Boolean	true	是否开启自动扫描功能,开启后将会自动扫描符合"components/组件名称/组件名称.vue"目录结构的组件
custom	Object		以正则方式自定义组件匹配规则。如果autoscan属性不能满足需求,可以使用custom自定义匹配规则

如果需要匹配 node_modules 内的 VUE 文件，需要使用"packageName/path/to/vue-file-$1.vue"形式的匹配规则，其中 packageName 为安装的包名，/path/to/vue-file-$1.vue 为 VUE 文件在包内的路径。自定义 easycom 配置的示例如下。

```
"easycom":{
  "autoscan":true,
  "custom":{
    "^uni-(.*)":"@/components/uni-$1.vue", // 匹配 components 目录下的 VUE 文件
    "^vue-file-(.*)":"packageName/path/to/vue-file-$1.vue" // 匹配 node_modules 内的 VUE 文件
  }
}
```

3. npm 方式

扩展组件也可以用 npm 方式来安装。uni-ui、uView 依赖于 scss，所以必须要安装此插件，否则组件无法正常运行。如果项目是使用 vue-cli 创建的，则需要安装 scss，并添加一个配置项。

安装步骤如下。

（1）安装 Sass，如果是 HBuilderX 项目，可以略过这个步骤。

```
npm i sass -D
npm i sass-loader@10 -D
```

（2）安装 uni-ui。

```
npm i @dcloudio/uni-ui
```

（3）配置 easycom。

使用 npm 方式安装好 uni-ui 之后，需要配置 easycom，让通过 npm 方式安装的组件支持 easycom。打开项目根目录下的 pages.json 并添加 easycom 节点。

```
{
    "easycom":{
        "autoscan":true,
        "custom":{
            // uni-ui 规则配置
            "^uni-(.*)":"@dcloudio/uni-ui/lib/uni-$1/uni-$1.vue"
        }
    },

    // 其他内容
    pages:[
      // …
    ]
}
```

（4）配置 vue.config.js，如果是 HBuilderX 项目，可以略过该步骤。

在根目录下创建 vue.config.js 文件，并进行如下配置。

```
module.exports = {
      transpileDependencies:['@dcloudio/uni-ui']
}
```

（5）使用组件。

```
<uni-badge text="1"></uni-badge>
```

注意：

也可以在新建项目时，直接新建一个 uni-ui 项目。

3.3.2 uni-scss 辅助样式

uni-ui 安装完成后，就可以在页面中使用 uni-ui 组件了。实例 3-8 介绍了 uni-ui 基本的使用方法。本小节将介绍更多的使用方法。

uni-ui 扩展组件中有很多依赖 uni-scss 辅助样式的组件。完整安装 uni-ui 或者安装依赖 uni-scss 的扩展组件时都会安装 uni-scss。在 uni-scss 中定义了一系列的 SCSS 文件。其中 index.scss 文件包含 SCSS 组件中所有的 SCSS。index.scss 文件通过编译可以生成.css 格式的文件。选择 uni-modules 目录下 uni-scss 文件夹下的 index.scss 文件，单击鼠标右键，选择【外部命令】—【scss/sass 编译】—【1 编译 scss/sass】，会在当前目录下生成 index.css 文件，如图 3-11 所示。

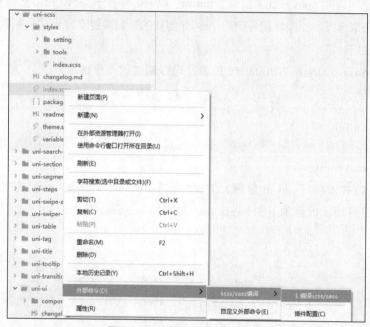

图 3-11　生成 index.css 文件

打开 index.css 文件，可以看到 uni-scss 提供的各种工具类。以下代码为 index.css 文件的前面几行。

```
@charset "UTF-8";
/* 水平间距 */
.uni-border {
  border:1px #F0F0F0 solid; }
.uni-primary {
  color:#2979ff; }
.uni-primary-bg {
  background-color:#2979ff; }
.uni-primary-disable {
  color:#8fb9ff; }
.uni-primary-disable-bg {
  background-color:#8fb9ff; }
.uni-primary-light {
```

```
  color:#a9c9ff; }
.uni-primary-light-bg {
  background-color:#a9c9ff; }
.uni-success {
  color:#18bc37; }
//省略
```

注意：

（1）如果没有编译工具，请先安装 scss 编译插件；

（2）生成 index.css 不是必需的，这里生成只是便于之后学习。

1．颜色类

前景色可以直接使用与变量同名的颜色类进行设置，对元素快速应用 color 样式。

```
<view class="uni-primary">主色</view>
<view class="uni-success">成功色</view>
<view class="uni-warning">警告色</view>
<view class="uni-error">错误色</view>
```

背景色可以在以上变量名的基础上加上 -bg 进行设置，对元素快速应用 background-color 样式。

```
<view class="uni-primary-bg">主色</view>
<view class="uni-success-bg">成功色</view>
<view class="uni-warning-bg">警告色</view>
<view class="uni-error-bg">错误色</view>
```

2．边框半径类

使用边框半径类可对元素快速应用 border-radius 样式。边框半径类的语法格式如下。

```
uni-radius-${direction}-${size}
```

direction 的取值：边框半径可以通过 t、r、b、l、tl、tr、br、bl 来设置 4 个角，其中 t 对应 top，r 对应 right，b 对应 bottom，l 对应 left。例如 uni-radius-t 、uni-radius-b-lg、uni-radius-tl 、uni-radius-br-lg。

size 的取值：基于 $uni-radius-root 变量的值，该变量的默认值为 5px，用于设置元素的 border-radius 属性。size 的具体取值如下。

- null：表示使用默认值（可忽略）。
- 0：表示清理所有圆角。
- sm：表示默认值 / 2（暂不支持）。
- lg：表示默认值×2。
- xl：表示默认值×6。
- pill：表示 9999px。
- circle：表示 50%（nvue 下不生效）。

例如：

```
<view class="uni-radius-circle"></view>    <!-- 边框为圆形 -->
<view class="uni-radius-t-0"></view>       <!-- 上面两个角为直角-->
```

3. 间距类

使用间距类可对元素快速应用 margin 或 padding 样式，范围是从 0 到 16。其语法格式如下。

```
uni-${property}${direction}-${size}
```

property：表示间距类型，其中 m 对应 margin，p 对应 padding。

direction：用于指定间距类所应用的边，其取值有 t、b、l、r、x、y、a。其中 t、b、l、r 分别代表 top、bottom、left、right，x 对应 left 和 right，y 对应 top 和 bottom，a 对应 all，表示所有方向。

size：取值范围为 0 到 16，以 4 为增量，即 1 代表 4px，2 代表 8px，以此类推。例如：

```
<view class="uni-mt-2"></view>     <!-- margin-top 为 8px -->
<view class="uni-mx-2"></view>     <!-- 左右 margin 为 8px-->
<view class="uni-py-5"></view>     <!-- 上下 padding 为 20px -->
```

4. 配置 scss

使用 uni-scss 时需在 App.vue 的 `<style lang='scss'></style>` 中引入 index.scss 文件。代码如下。

```
<style lang="scss">
  /*每个页面公共 CSS */
  @import '@/uni_modules/uni-scss/index.scss';
</style>
```

如果需要使用或修改 uni-scss 中的 SCSS 变量的值，则需要在项目的根目录的 uni.scss 文件中引入变量文件 variables.scss，示例如下。

```
/* 需要放到文件最前面 */
@import '@/uni_modules/uni-scss/variables.scss';
```

【实例 3-9】演示 uni-scss 辅助样式。

在本例中，使用颜色、边框半径、间距等工具类。

实例 3-9

实现步骤

（1）在 uniappch03 项目中新建页面 testui.vue 并生成同名文件夹，然后在 pages.json 中的 pages 节点进行注册。代码如下。

```
"pages":[{
        "path":"pages/index/index",
        "style":{
            "navigationBarTitleText":"uni-app",
            "enablePullDownRefresh":true
        }
    }, {
        "path":"pages/index/testload",
        "style":{
            "navigationBarTitleText":"uni-app"
        }
    }, {
        "path":"pages/testui/testui",
```

```
        "style":{
            "navigationBarTitleText":"uni-ui的演示",
            "enablePullDownRefresh":false
        }
    }]
```

(2)准备两组图标,将 index.vue 和 testui.vue 设置为 tabBar 页面。在 pages.json 文件中,添加 tabBar 节点(此过程参考第 2 章)。代码如下。

```
"tabBar":{
    "backgroundColor":"#F8F8F8",
    "color":"#8F8F94",
    "height":"60px",
    "list":[{
            "text":"生命周期",
            "pagePath":"pages/index/index",
            "iconPath":"static/life.png",
            "selectedIconPath":"static/lift-selected.png"
        }, {
            "text":"扩展组件",
            "pagePath":"pages/testui/testui",
            "iconPath":"static/ex-com.png",
            "selectedIconPath":"static/ex-com-selected.png"

        }]
    },
```

(3)在 testui.vue 中添加如下代码(需先在 App.vue 中引用 index.scss 文件)。

```
<template>
    <view class="content">
        <view class=" box uni-primary-bg  uni-mb-6">
            <text class="">人生不是要超越别人,而是要超越自己</text>
        </view>
</view>
</template>
<style lang="scss">
    .content {
        display:flex;
        flex-direction:column;
        align-items:center;
        justify-content:center;
    }
    .box {
        width:100%;
        height:100px;
        display:flex;
        justify-content:center;
        align-items:center;
        color:#FFFFFF;
        font-size:20px;
    }
</style>
```

（4）运行项目，查看 testui.vue 页面，效果如图 3-12 所示。

3.3.3 使用组件

图 3-12 uni-scss 辅助样式演示效果

uni-ui 中提供了一系列的扩展组件，作为基础组件的补充，如布局、表单、分组、折叠、日历、选择器等组件。本小节的实例 3-10 讲解如何使用相应组件，其中涉及 uni-card（卡片）组件以及布局组件中的 uni-row（行）和 uni-col（列）。

1. 流式栅格系统

uni-row、uni-col 组成流式栅格系统，将屏幕或视口分为 24 份，可以使用户迅速、简便地创建布局。每个 uni-row 分为 24 份，在 uni-row 里面应用 uni-col。每一个 uni-col 占一份或多份，一个 uni-row 中所有 uni-col 的总份数为 24，若超过则换到下一行。

uni-row 的属性只有一个 gutter，用于设置行与行的间距，默认值为 0。

uni-col 的属性如表 3-7 所示。

表 3-7 uni-col 的属性

属性	类型	说明
span	Number	栅格占据的列数，默认值为 24
offset	Number	栅格左侧间隔格数
push	Number	栅格向右偏移格数
pull	Number	栅格向左偏移格数
xs	Number/Object	屏幕宽度<768px 时，要显示的栅格规则，如：xs="8"或:xs="{span: 8, pull: 4}"
sm	Number/Object	屏幕宽度≥768px 时，要显示的栅格规则
md	Number/Object	屏幕宽度≥992px 时，要显示的栅格规则
lg	Number/Object	屏幕宽度≥1200px 时，要显示的栅格规则
xl	Number/Object	屏幕宽度≥1920px 时，要显示的栅格规则

2. uni-card 组件

uni-card 组件通常用来显示完整独立的一段信息，同时让用户理解它的作用。例如一篇文章的预览图、作者信息、时间等，卡片通常是更复杂和更详细信息的入口点。在 uni-card 组件的官网中打开组件下载页面，在下载页面中单击右侧的【使用 HBuilderX 导入示例项目】按钮，参考图 3-7，将导入组件的完整示例代码。以下为示例代码的部分。

```
<uni-section title="双标题卡片" type="line">
    <uni-card title="基础卡片" sub-title="副标题" extra="额外信息" :thumbnail="avatar" @click="onClick">
        <text class="uni-body">这是一个带头像和双标题的基础卡片,此示例展示了一个完整的卡片。</text>
    </uni-card>
</uni-section>
```

uni-card 组件的演示效果如图 3-13 所示。

图 3-13 uni-card 组件的演示效果

uni-card 组件的属性如表 3-8 所示。

表 3-8 uni-card 组件的属性

属性	类型	默认值	说明
title	String		标题文字
sub-title	String		副标题文字
extra	String		标题额外信息
thumbnail	String		标题左侧缩略图,支持网络图片、本地图片,本地图片需要传入绝对路径,如/static/xxx.png
cover	String		封面图,支持网络图片、本地图片,本地图片需要传入绝对路径,如/static/xxx.png
is-full	Boolean	false	内容是否通栏,为 true 时将去除 padding 值
is-shadow	Boolean	false	卡片内容是否开启阴影
shadow	String	默认值见说明部分	卡片阴影,需符合 CSS 值,默认值为 0px 0px 3px 1px rgba(0, 0, 0, 0.08)
border	Boolean	true	卡片边框
margin	String	10px	卡片外边距
spacing	String	10px	卡片内边距
padding	String	10px	卡片内容内边距

【实例 3-10】演示 uni-card 的使用。

本例中使用 uni-card 的基本形式。卡片中的内容使用 uni-row 和 uni-col 来进行布局。

实例 3-10

<template></template>标签内的代码如下。

```
<template>
    <view class="content">
        ……省略实例 3-9 的代码
        <uni-card class="uni-card1  ">
            <uni-row>
                <uni-col :span="4">
                    <image    class=""   src="../../static/images/icon/ide.png" mode="widthFix" style="width:100%; ">
                    </image>
                </uni-col>
                <uni-col :span="20">
```

```
                    <view class="uni-ml-4">
                        <text class="title">开发工具</text>
                        <view>Eclipse、IntelliJ IDEA、Visual Studio Code、Sublime、HBuilder</view>
                    </view>
                </uni-col>
            </uni-row>
        </uni-card>
        ……省略了其他3个类似卡片
    </view>
</template>
```

<style></style>标签内的代码如下。

```
.title {
    font-size:24px;
    font-weight:400;
    color:#555555;
}
.uni-card1 {
    width:90%;
    margin-top:0 !important;
}
```

运行 uniappch03 项目,选择扩展组件 testui.vue 页面,页面效果如图 3-14 所示。

3. uni-list 组件

uni-list(列表)组件包含基本列表样式,可扩展插槽,具有长列表性能优化、多平台兼容的特点。uni-list 组件是父容器,其核心是 uni-list-item 子组件,它代表列表中的一个可重复行,子组件可以无限循环。其关联组件有 uni-list-item、uni-badge、uni-icons、uni-list-chat。

uni-list-item 有很多样式,可通过内置的属性用于一些常见的场景。当内置属性不满足需求时,可以通过扩展插槽来自定义列表内容。

内置属性适用的场景包括导航列表、设置列表、小图标列表、通信录列表、聊天记录列表。

涉及有很多大图或丰富内容的列表时,比如类似今日头条的新闻列表、类似淘宝的电商列表,需要通过扩展插槽实现。

uni-list-item 的常用属性如表 3-9 所示。

图 3-14 uni-card 演示效果

表 3-9 uni-list-item 的常用属性

属性	类型	默认值	说明
title	String		标题
note	String		描述
ellipsis	Number	0	title 是否溢出隐藏,可选值:0 表示默认;1 表示显示一行;2 表示显示两行

续表

属性	类型	默认值	说明
thumb	String		左侧缩略图，若 thumb 有值，则不会显示扩展图标
thumbSize	String	medium	缩略图尺寸，可选值：lg 表示大图；medium 表示一般；sm 表示小图
showBadge	Boolean	false	是否显示数字角标
badgeText	String		数字角标内容
badgeType	String		数字角标类型，参考 uni-app 官网的 uni-icons 图标组件
badgeStyle	Object		数字角标样式，使用 uni-app 官网的 uni-badge 的徽章组件的 custom-style 参数
rightText	String		右侧文字内容
disabled	Boolean	false	是否禁用
showArrow	Boolean	true	是否显示箭头图标
link	String	navigateTo	新页面跳转方式，可选值有 redirectTo、reLaunch、switchTab、navigateTo
to	String		新页面跳转地址，如填写此属性，click 事件会返回页面是否跳转成功

uni-list-item 事件如表 3-10 所示。

表 3-10　uni-list-item 事件

事件	说明	返回参数
click	点击 uniListItem 触发事件，需开启点击反馈	
switchChange	点击切换 Switch 时触发，需显示 Switch	e={value:checked}

3.4　案例：新闻列表页和新闻详情页

本案例利用本章知识完成新闻列表页（见图 3-15）和新闻详情页（见图 3-16）。

图 3-15　新闻列表页

图 3-16　新闻详情页

案例：新闻列表页和新闻详情页

实现步骤

（1）新建 uni-app 项目 ch03_news，然后按照 3.3.3 小节的内容导入 uni-list 组件及其关联组件。

（2）在 index.vue 页面中添加如下代码。

```
<template>
    <view class="container">
        <uni-list>
            <uni-list-item  v-for="(item,i) in news" :key="i"   :title="item.title" :note="item.author_name" showArrow
                :thumb="item.cover"
                thumb-size="lg"  link="navigateTo"  :to="'../info/info?post_id='+item.post_id" />
        </uni-list>
    </view>
</template>
<script>
    export default {
        data() {
            return {
                news:[]
            }
        },
        onLoad() {
            uni.request({
                url:'https://unidemo.dcloud.net.cn/api/news',
                method:'GET',
                data:{},
                success:res => {
                    this.news = res.data;
                    console.log(this.news);
                },
                fail:() => {},
                complete:() => {}
            })
        },
        methods:{          }
    }
</script>
<style>
    .container {
        padding:20px;
        font-size:14px;
        line-height:24px;
    }
</style>
```

在 onLoad 函数中，调用网络请求 API 函数 uni.request 访问远程服务器，得到数据，并把数据赋给 news 数组。数组中每一项数据的格式如图 3-17 所示。

在 <template></template> 标签内，使用 uni-list 显示数组 news。点击列表项，跳转到新闻详情页，跳转时带上 id 参数 ":to="'../info/info?post_id='+item.post_id""。

```
"author_avatar": "https://      .com/avatar/201906/14092708/q42867125b5hasjfl120",
"author_email": "",
"author_name": "神译局",
"column_id": "103",
"column_name": "技能Get",
"comments_count": 0,
"content": "<blockquote><p>神译局是36氪旗下编译团队，关注科技、商业、职场、生活等领域，重点介
"cover": "https://      .com/20200406/v2_d2c6a686b4074a1eb43603e67d6ba204_img_png",
"created_at": "2020-04-11 13:52:12",
"from_id": "36kr",
"id": 121367,
"post_id": "5309137",
"published_at": "2020-04-11 13:47:01",
"store_at": "0000-00-00 00:00:00",
"summary": "学会利用好真正艰难的时刻，增强自己的精神力量",
"title": "克服危机和压力，精神力量强大的人都有这5个习惯",
"type": "news",
"updated_at": "2020-04-11 13:52:12",
"views_count": 0
```

图 3-17　返回数据格式

（3）新建一个页面 info.vue，通过 info.vue 的 onLoad 访问并显示对应的新闻详情页。具体代码如下。

```
<template>
    <view class="container">
        <view class="title"><text>{{title}}</text></view>
        <view>
            <rich-text :nodes="contents"></rich-text>
        </view>
    </view>
</template>
<script>
    export default {
        data() {
            return {
                title:"",
                contents:""
            }
        },
        onLoad(op) {
            console.log(op);
            uni.request({
                url:'https://unidemo.dcloud.net.cn/api/news/36kr/'+op.post_id,
                method:'GET',
                data:{},
                success:res => {
                    console.log(res);
                    this.title = res.data.title;
                    this.contents = res.data.content;
                },
                fail:() => {},
                complete:() => {}
            });
        },
        methods:{         }
    }
</script>
<style>
    .container {
```

```
            padding:20px;
            font-size:14px;
            line-height:24px;
        }
        .title{
            font-weight:700;
            font-size:1.5rem;
            line-height:2.5rem;
        }
</style>
```

（4）运行项目查看效果。

本章小结

本章主要介绍了 uni-app 的生命周期、条件编译、扩展组件 uni-ui 的安装和使用等。最后通过新闻列表页和新闻详情页案例综合应用了相关知识点。

项目实战

完成小程序应用中的个人中心页，页面效果如图 3-18 所示，可以使用 uni-ui 或 uView 等扩展组件。

图 3-18　个人中心页

拓展实训项目

国家反诈中心 App 是国务院打击治理电信网络新型违法犯罪工作部际联席会议合成作战平台，集资源整合、情报研判、侦查指挥为一体，在打击、防范、治理电信网络诈骗等新型违法犯罪中发挥着重要作用。请安装该 App，然后利用本章所学的扩展组件实现其中的页面。

第4章
uni-app组件

本章导读

本章主要讲解 uni-app 的常用组件，包括容器组件、基础组件、表单组件、媒体组件、地图组件等。通过对本章的学习，读者能够使用 uni-app 的常用组件进行开发。

学习目标

知识目标	1. 掌握 uni-app 的常用组件的属性 2. 掌握 uni-app 的常用组件的事件
能力目标	1. 具有灵活使用常用组件搭建页面的能力 2. 具有熟练使用组件的常用属性的能力
素质目标	1. 具有良好的软件编码规范素养 2. 培养技能报国的爱国主义情怀、精益求精的工匠精神 3. 具有较强的自主学习与钻研精神

知识思维导图

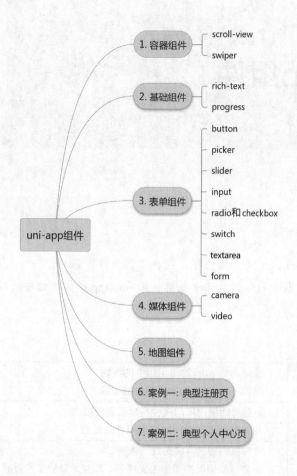

4.1 容器组件

uni-app 为开发者提供了一系列容器组件，包括 view、swiper 等，它们本身不显示任何可视化元素。它们的用途都是包裹其他真正用于显示的组件。常用的容器组件有 view、swiper、scroll-view 等，其中 view 组件参见 2.5.1 小节。本节介绍 scroll-view 组件和 swiper 组件。

4.1.1 scroll-view

scroll-view（可滚动视图）组件用于实现区域滚动，其效果类似于 H5 的 iScroll。注意，在 WebView 渲染的页面中，区域滚动的性能不及页面滚动性能。

scroll-view 使用场景非常多，如内容溢出时滚动显示内容、横向滚动、纵向滚动、下拉刷新、上拉加载等。scroll-view 的属性及说明如表 4-1 所示。

表4-1　scroll-view 的属性及说明

属性	类型	默认值	说明
scroll-x	Boolean	false	允许横向滚动
scroll-y	Boolean	false	允许纵向滚动
upper-threshold	Number/String	50	距顶部/左边多远时（单位为 px），触发 scrolltoupper 事件
lower-threshold	Number/String	50	距底部/右边多远时（单位为 px），触发 scrolltolower 事件
scroll-top	Number/String		设置纵向滚动条位置
scroll-left	Number/String		设置横向滚动条位置
scroll-into-view	String		其值应为某子元素 ID（ID 不能以数字开头）。设置哪个方向可滚动，则沿哪个方向滚动到该元素
scroll-with-animation	Boolean	false	在设置滚动条位置时使用动画过渡
@scrolltoupper	EventHandle		滚动到顶部/左边，触发 scrolltoupper 事件
@scrolltolower	EventHandle		滚动到底部/右边，触发 scrolltolower 事件
@scroll	EventHandle		滚动时触发 scroll 事件，如 event.detail = {scrollLeft, scrollTop, scrollHeight, scrollWidth, deltaX, deltaY}
enable-back-to-top	Boolean	false	当点击顶部状态栏（iOS）或双击标题栏（Android）时，滚动条返回顶部，只支持竖向
show-scrollbar	Boolean	false	控制是否出现滚动条
refresher-enabled	Boolean	false	开启自定义下拉刷新功能
refresher-threshold	Number	45	设置自定义下拉刷新阈值
refresher-default-style	String	"black"	设置自定义下拉刷新默认样式，支持设置 black、white、none，none 表示不使用默认样式
refresher-background	String	"#FFF"	设置自定义下拉刷新区域背景颜色
refresher-triggered	Boolean	false	设置当前下拉刷新状态，true 表示下拉刷新已经被触发，false 表示下拉刷新未被触发
enable-flex	Boolean	false	启用 flex 布局。启用后，当前节点声明的 display: flex 就会成为 flex container，并作用于其子节点
scroll-anchoring	Boolean	false	开启 scroll anchoring 特性，即控制滚动位置不随内容变化而抖动，仅在 iOS 下生效，在 Android 下可使用 CSS 的 overflow-anchor 属性
@refresherpulling	EventHandle		自定义下拉刷新控件被下拉
@refresherrefresh	EventHandle		自定义下拉刷新控件被触发
@refresherrestore	EventHandle		自定义下拉刷新控件被复位
@refresherabort	EventHandle		自定义下拉刷新控件被中止

平台差异说明： 以上属性中，前面 11 个属性各平台均支持，后面的属性有平台限制，一般情况下微信小程序均支持，其他小程序不支持。具体细节请读者参见 uni-app 官网相关说明。

使用纵向滚动时，需要给 scroll-view 设置一个固定高度，通过 CSS 的 height 实现；使用横向滚动时，需要给 scroll-view 添加 white-space: nowrap;样式。

【实例 4-1】演示 scroll-view 的使用。

实例 4-1

实现步骤

（1）使用默认模板新建一个 uni-app 项目 uniappch04，本章实例将使用该项目。

（2）在项目中新建一个名为"scroll-view"的页面并勾选【在 pages.json 中注册】。

（3）在 pages.json 文件中将刚创建页面的 navigationBarTitleText 设为 scroll-view。

（4）在 scroll-view.vue 中完成如下代码的编写。

```
<template>
    <view>
        <view class="uni-padding-wrap uni-common-mt">
            <view class="uni-title uni-common-mt">
                Vertical Scroll
                <text>\n 纵向滚动</text>
            </view>
            <view>
                <scroll-view :scroll-top="scrollTop" scroll-y="true" class="scroll-Y" @scrolltoupper="upper"
                    @scrolltolower="lower" @scroll="scroll">
                    <view id="demo1" class="scroll-view-item uni-bg-red">富强 民主 文明 和谐</view>
                    <view id="demo2" class="scroll-view-item uni-bg-green">自由 平等 公正 法治</view>
                    <view id="demo3" class="scroll-view-item uni-bg-blue">爱国 敬业 诚信 友善</view>
                </scroll-view>
            </view>
            <view @tap="goTop" class="uni-link uni-center uni-common-mt">
                点击这里返回顶部
            </view>

            <view class="uni-title uni-common-mt">
                Horizontal Scroll
                <text>\n 横向滚动</text>
            </view>
            <view>
                <scroll-view class="scroll-view_H" scroll-x="true" @scroll="scroll" scroll-left="120">
                    <view id="demo1" class="scroll-view-item_H uni-bg-red">富强 民主 文明 和谐</view>
                    <view id="demo2" class="scroll-view-item_H uni-bg-green">自由 平等 公正 法治</view>
```

```html
                    <view id="demo3" class="scroll-view-item_H uni-bg-blue">爱
国 敬业 诚信 友善</view>
                </scroll-view>
            </view>
            <view class="uni-common-pb"></view>
        </view>
    </view>
</template>
<script>
    export default {
        data() {
            return {
                scrollTop:0,
                old:{
                    scrollTop:0
                }
            }
        },
        methods:{
            upper:function(e) {
                console.log(e)
            },
            lower:function(e) {
                console.log(e)
            },
            scroll:function(e) {
                console.log(e)
                this.old.scrollTop = e.detail.scrollTop
            },
            goTop:function(e) {
                // 解决 view 层不同步的问题
                this.scrollTop = this.old.scrollTop
                this.$nextTick(function() {
                    this.scrollTop = 0
                });
                uni.showToast({
                    icon:"none",
                    title:"纵向滚动，scrollTop 值已被修改为 0"
                })
            }
        }
    }
</script>
<style>
    .scroll-Y {
        height:300rpx;
    }
    .scroll-view_H {
        white-space:nowrap;
        width:100%;
    }
    .scroll-view-item {
        height:300rpx;
```

```
            line-height:300rpx;
            text-align:center;
            font-size:36rpx;
        }
        .scroll-view-item_H {
            display:inline-block;
            width:100%;
            height:300rpx;
            line-height:300rpx;
            text-align:center;
            font-size:36rpx;
        }
        .uni-bg-red{
            background:#F76260; color:#FFF;
        }
    .uni-bg-green{
            background:#09BB07; color:#FFF;
        }
        .uni-bg-blue{
            background:#007AFF; color:#FFF;
        }
</style>
```

（5）在当前页面选择【运行】—【运行到浏览器】，即可查看运行效果，如图 4-1 所示。

4.1.2 swiper

swiper（滑块）组件的最常用的功能之一就是制作轮播图效果，可以兼容多个平台。

swiper 一般用于左右或上下滑动，如制作 banner 轮播图。注意，滑动切换是一屏一屏地切换。swiper 下的每一个 swiper-item 都是一个滑动切换区域，不能停留在两个滑动切换区域之间。swiper 的属性及说明如表 4-2 所示。

图 4-1 scroll-view 演示效果

表 4-2 swiper 的属性及说明

属性	类型	默认值	说明
indicator-dots	Boolean	false	是否显示面板指示点
indicator-color	Color	rgba(0, 0, 0, .3)	指示点颜色
indicator-active-color	Color	#000000	当前选中的指示点颜色
autoplay	Boolean	false	是否自动切换
current	Number	0	当前所在滑块的索引
current-item-id	String		当前所在滑块的 item-id，不能与 current 同时指定

续表

属性	类型	默认值	说明
interval	Number	5000	自动切换时间间隔
duration	Number	500	滑动动画时长
circular	Boolean	false	是否采用衔接滑动，即播放到末尾后重新回到开头
vertical	Boolean	false	滑动方向是否为纵向
previous-margin	String	0px	前边距，可用于露出前一项的一小部分，接收以 px 和 rpx 为单位的值
next-margin	String	0px	后边距，可用于露出后一项的一小部分，接收以 px 和 rpx 为单位的值
display-multiple-items	Number	1	同时显示的滑块数量
skip-hidden-item-layout	Boolean	false	是否跳过未显示的滑块布局，若设为 true 可优化复杂情况下的滑动性能，但会丢失隐藏状态下滑块的布局信息
disable-touch	Boolean	false	是否禁止用户触摸操作
touchable	Boolean	true	是否监听用户的触摸事件，只在初始化时有效，不能动态变更
easing-function	String	default	指定 swiper 切换缓动动画类型，有效值：default、linear、easeInCubic、easeOutCubic、easeInOutCubic
@change	EventHandle		current 改变时会触发 change 事件，event.detail = {current: current, source: source}
@transition	EventHandle		swiper-item 的位置发生改变时会触发 transition 事件，event.detail = {dx: dx, dy: dy}，支付宝小程序暂不支持 dx、dy
@animationfinish	EventHandle		动画结束时会触发 animationfinish 事件，event.detail = {current: current, source: source}

平台差异说明：表 4-2 中列举的属性多数平台都支持。另外，支付宝小程序还支持 active-class、changing-class、acceleration、disable-programmatic-animation。具体细节请读者参考 uni-app 官网。

swiper 使用起来很简单，但要注意 swiper-item 内部的 image 组件并非 HTML 元素，而是 uni-app 内置的 image 组件。在 uni-app 中不要使用 img 元素而要使用 image 组件装载图片，image 组件内容参见 2.5.4 小节。

【实例 4-2】演示 swiper 的使用。

⚙ **实现步骤**

（1）在项目 uniappch04 中新建一个名为"swiper"的页面并勾选【在 pages.json 中注册】。

（2）在 pages.json 文件中将刚创建页面的 navigationBarTitleText 设为 swiper。

实例 4-2

（3）在 swiper.vue 中完成如下代码的编写。

```html
<template>
    <view>
        <view class="uni-margin-wrap">
            <swiper class="swiper"
                circular :indicator-dots="indicatorDots" :autoplay="autoplay"
                :interval="interval"
                :duration="duration">
                <swiper-item>
                    <view class="swiper-item uni-bg-red">人民有信仰</view>
                </swiper-item>
                <swiper-item>
                    <view class="swiper-item uni-bg-green">国家有力量</view>
                </swiper-item>
                <swiper-item>
                    <view class="swiper-item uni-bg-blue">民族有希望</view>
                </swiper-item>
            </swiper>
        </view>
        <view class="swiper-list">
            <view class="uni-list-cell uni-list-cell-pd">
                <view class="uni-list-cell-db">指示点</view>
                <switch :checked="indicatorDots" @change="changeIndicatorDots" />
            </view>
            <view class="uni-list-cell uni-list-cell-pd">
                <view class="uni-list-cell-db">自动播放</view>
                <switch :checked="autoplay" @change="changeAutoplay" />
            </view>
        </view>
        <view class="uni-padding-wrap">
            <view class="uni-common-mt">
                <text>幻灯片切换时长（ms）</text>
                <text class="info">{{duration}}</text>
            </view>
            <slider @change="durationChange" :value="duration" min="500" max="2000"/>
            <view class="uni-common-mt">
                <text>自动播放间隔时长（ms）</text>
                <text class="info">{{interval}}</text>
            </view>
            <slider @change="intervalChange" :value="interval" min="2000" max="10000"/>
        </view>
    </view>
</template>
<script>
export default {
    data() {
        return {
            background:['color1', 'color2', 'color3'],
            indicatorDots:true,
            autoplay:true,
            interval:2000,
```

```
                    duration:500
            }
        },
        methods:{
            changeIndicatorDots(e) {
                this.indicatorDots = !this.indicatorDots
            },
            changeAutoplay(e) {
                this.autoplay = !this.autoplay
            },
            intervalChange(e) {
                this.interval = e.target.value
            },
            durationChange(e) {
                this.duration = e.target.value
            }
        }
    }
</script>
<style>
    .uni-margin-wrap {
        width:690rpx;
        width:100%;
    }
    .swiper {
        height:300rpx;
    }
    .swiper-item {
        display:block;
        height:300rpx;
        line-height:300rpx;
        text-align:center;
    }
    .swiper-list {
        margin-top:40rpx;
        margin-bottom:0;
    }
    .uni-common-mt {
        margin-top:60rpx;
        position:relative;
    }
    .info {
        position:absolute;
        right:20rpx;
    }
    .uni-padding-wrap {
        width:550rpx;
        padding:0 100rpx;
    }
    /* 背景色 */
    .uni-bg-red{
        background:#F76260; color:#FFF;
    }
    .uni-bg-green{
```

```
        background:#09BB07; color:#FFF;
    }
    .uni-bg-blue{
        background:#007AFF; color:#FFF;
    }
    /* 列表 */
    .uni-list-cell {
        position:relative;
        display:flex;
        flex-direction:row;
        justify-content:space-between;
        align-items:center;
    }
    .uni-list-cell-pd {
        padding:22rpx 30rpx;
    }
    .uni-list-cell-db,
    .uni-list-cell-right {
        flex:1;
    }
    .uni-list-cell::after {
        position:absolute;
        z-index:3;
        right:0;
        bottom:0;
        left:30rpx;
        height:1px;
        content:'';
        -webkit-transform:scaleY(.5);
        transform:scaleY(.5);
        background-color:#c8c7cc;
    }
</style>
```

（4）在当前页面选择【运行】—【运行到浏览器】，即可查看运行效果，如图 4-2 所示。

图 4-2 swiper 演示效果

4.2 基础组件

基础组件包括 icon 组件、text 组件、rich-text 组件、progress 组件，其中常用组件有 text、rich-text、progress，text 组件内容参见 2.5.2 小节。本节介绍 rich-text 组件和 progress 组件。

4.2.1 rich-text

rich-text（富文本）组件可以解析 HTML 标签，通常用于显示商品介绍信息、文章内容等。rich-text 的属性说明如表 4-3 所示。

表 4-3 rich-text 的属性说明

属性	类型	默认值	说明
space	String		显示连续空格
nodes	Array / String	[]	节点列表 / HTML String
selectable	Boolean	true	富文本是否可以长按选中，可用于复制、粘贴等场景
image-menu-prevent	Boolean	false	是否阻止长按图片时弹出默认菜单。如需阻止，就设置 image-menu-prevent 属性的值为 true，否则将不设置该属性。该属性只在初始化时有效，不能动态变更
preview	Boolean		富文本中的图片是否可点击预览。在不设置该属性的情况下，若 rich-text 未监听点击事件，则默认开启点击预览功能。未显式设置 preview 时会进行默认点击预览判断，建议显式设置 preview
@itemclick	EventHandle		拦截点击事件（只支持<a></a>、<img></img>标签），返回当前节点信息，event.detail={node}

平台差异说明：selectable、image-menu-prevent、preview 仅百度小程序支持，@itemclick 为 H5 平台和 App 平台支持。

【实例 4-3】演示 rich-text 的使用。

⚙ **实现步骤**

（1）在项目 uniappch04 中新建一个名为 "rich-text" 的页面并勾选【在 pages.json 中注册】。

实例 4-3

（2）在 pages.json 文件中将刚创建页面的 navigationBarTitleText 设为 rich-text。

（3）在 rich-text.vue 中完成如下代码的编写。

```
<template>
    <view class="content">
        <page-head :title="title"></page-head>
        <view class="uni-padding-wrap">
            <view class="uni-title uni-common-mt">
                数组类型
                <text>\nnodes 属性为 Array</text>
```

```html
        </view>
        <view class="uni-common-mt" style="background:#FFF; padding:20rpx;">
            <rich-text :nodes="nodes"></rich-text>
        </view>
        <view class="uni-title uni-common-mt">
            字符串类型
            <text>\nnodes 属性为 String</text>
        </view>
        <view class="uni-common-mt" style="background:#FFF; padding:20rpx;">
            <rich-text :nodes="strings"></rich-text>
        </view>
      </view>
    </view>
</template>
<script>
export default {
    data() {
        return {
            nodes:[{
                name:'div',
                attrs:{
                    class:'div-class',
                    style:'line-height:60px; color:red; text-align:center;'
                },
                children:[{
                    type:'text',
                    text:'新时代 新征程 争出彩!'
                }]
            }],
            strings:'<div style="text-align:center;"><img src="../../static/danghui.jpeg""/></div>'
        }
    }
}
</script>
<style>
    img{
        height:72px;
        width:72px;
    }
    .uni-padding-wrap{
        /* width:690rpx; */
        padding:0 30rpx;
    }
    .uni-title {
        font-size:30rpx;
        font-weight:500;
        padding:20rpx 0;
        line-height:1.5;
    }
    .uni-common-mt{
        margin-top:30rpx;
```

（4）在当前页面选择【运行】—【运行到浏览器】，即可查看运行效果，如图 4-3 所示。

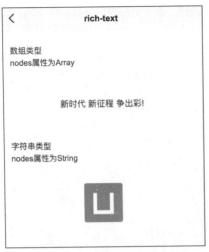

图 4-3　rich-text 演示效果

4.2.2　progress

progress（进度条）组件是一种可以提高用户体验的组件，用于显示内容加载或操作处理进度。progress 的属性说明如表 4-4 所示。

表 4-4　progress 的属性说明

属性	类型	默认值	说明
percent	Number		百分比，取值范围为 0～100
show-info	Boolean	false	是否在进度条右侧显示百分比
stroke-width	Number	6	进度条的宽度，单位为 px
activeColor	Color	#09BB07	已选择的进度条的颜色
backgroundColor	Color	#EBEBEB	未选择的进度条的颜色
active	Boolean	false	进度条从左往右的动画
border-radius	Number/String	0	圆角大小
font-size	Number/String	16	进度条右侧百分比的字号大小
active-mode	String	backwards	backwards：动画从头播放。forwards：动画从上次的结束点接着播放
duration	Number	30	进度增加 1% 所需时间（单位为 ms）
@activeend	EventHandle		动画完成事件

平台差异说明：前面 6 个属性各平台均支持，其他主要由 App、微信、QQ、京东、快手小程序支持。具体细节参考 uni-app 官网。

【实例 4-4】演示 progress 的使用。

实现步骤

（1）在项目 uniappch04 中新建一个名为"progress"的页面并勾选【在 pages.json 中注册】。

实例 4-4

（2）在 pages.json 文件中将刚创建页面的 navigationBarTitleText 设为 progress。

（3）在 progress.vue 中完成如下代码的编写。

```vue
<template>
    <view>
        <view class="uni-padding-wrap uni-common-mt">
            <view class="progress-box">
                <progress :percent="pgList[0]" show-info stroke-width="3" />
            </view>
            <view class="progress-box">
                <progress :percent="pgList[1]" stroke-width="3" />
            </view>
            <view class="progress-box">
                <progress :percent="pgList[2]" activeColor="#10AEFF" stroke-width="3"/>
            </view>
            <view class="progress-control">
                <button type="primary" @click="setProgress">设置进度</button>
                <button type="warn" @click="clearProgress">清除进度</button>
            </view>
        </view>
    </view>
</template>
<script>
    export default {
        data() {
            return {
                pgList:[0, 0, 0]
            }
        },
        methods:{
            setProgress() {
                this.pgList = [30, 60, 90]
            },
            clearProgress() {
                this.pgList = [0, 0, 0]
            }
        }
    }
</script>
<style>
    progress{
        width:100%;
    }
    .progress-box {
        display:flex;
        height:50rpx;
```

```
        margin-bottom:60rpx;
    }
    .uni-icon {
        line-height:1.5;
    }
    .progress-cancel {
        margin-left:40rpx;
    }
    .progress-control button{
        margin-top:20rpx;
    }
    .uni-padding-wrap{
        padding:0 30rpx;
    }
    .uni-common-mt{
        margin-top:30rpx;
    }
</style>
```

（4）在当前页面选择【运行】—【运行到浏览器】，即可查看运行效果，如图 4-4 所示。

图 4-4　progress 演示效果

4.3　表单组件

uni-app 提供了丰富的表单组件，包括 button 组件、picker 组件、slider 组件、input 组件、radio 组件、checkbox 组件、switch 组件、textarea 组件、form 组件等，本节将介绍这些组件的使用。

4.3.1　button

uni-app 的 button（按钮）组件与 H5 的<button></button>标签的功能有所不同，其功能非常强大，属性有很多。button 属性说明如表 4-5～表 4-8 所示。

表 4-5　button 的属性说明

属性	类型	默认值	说明
size	String	default	按钮的大小
type	String	default	按钮的样式类型
plain	Boolean	false	按钮是否为镂空，背景透明

续表

属性	类型	默认值	说明
disabled	Boolean	false	是否禁用按钮
loading	Boolean	false	按钮名称前是否带 loading 图标
form-type	String		用于 form 组件，点击按钮分别会触发 form 组件的 submit、reset 事件
open-type	String		描述按钮的开放能力，具体取值见表 4-8
hover-class	String	button-hover	指定按钮按下时的样式类。当 hover-class="none" 时，没有点击态效果
hover-start-time	Number	20	按钮按住后多久出现点击态，单位 ms
hover-stay-time	Number	70	手指松开后点击态保留时间，单位 ms
app-parameter	String		打开 App 时，向 App 传递的参数，open-type=launchApp 时有效
hover-stop-propagation	Boolean	false	指定是否阻止某节点的祖先节点出现点击态
lang	String	'en'	指定返回用户信息的语言，zh_CN 表示简体中文，zh_TW 表示繁体中文，en 表示英文

平台差异说明：在以上属性中，app-parameter 属性被 QQ、微信小程序支持，hover-stop-propagation、lang 属性仅被微信小程序支持。除此之外，当 open-type 属性为指定值时，对应平台支持的属性由于篇幅限制这里没有列举，具体请读者参考 uni-app 官网文档。例如，当 open-type="openGroupProfile"时，QQ 小程序支持的属性有 groud-id，表示打开群资料卡时，传递的群号。

表 4-6　size 属性说明

值	说明
default	默认尺寸
mini	小尺寸

表 4-7　type 属性说明

值	说明
primary	微信小程序、360 小程序为绿色，App、H5、百度小程序、支付宝小程序、飞书小程序、快应用为蓝色，字节跳动小程序为红色，QQ 小程序为浅蓝色。如想在多平台统一颜色，请使用 default，然后自行编写样式
default	白色
warn	红色

表 4-8　open-type 属性说明

值	说明
feedback	打开意见反馈页面，用户可提交反馈内容并上传日志
share	触发用户转发操作

续表

值	说明
getUserInfo	获取用户信息,可以从@getuserinfo 回调中获取用户信息
contact	打开客服会话,如果用户在会话中点击消息卡片后返回应用,可以从@contact 回调中获得具体信息
getPhoneNumber	获取用户手机号,可以从@getphonenumber 回调中获取用户手机号
launchApp	在小程序中打开 App,可以通过 app-parameter 属性设定向 App 传递的参数
openSetting	打开授权设置页
chooseAvatar	获取用户头像,可以从@chooseavatar 回调中获取用户头像信息
uploadDouyinVideo	发布抖音视频
getAuthorize	支持小程序授权
lifestyle	关注生活号
contactShare	分享到通信录好友
openGroupProfile	呼起 QQ 群资料卡页,可以通过 group-id 属性设定需要打开的群资料卡的群号,同时在 manifest.json 中必须配置 groupIdList
openGuildProfile	呼起频道页,可以通过 guild-id 属性设定需要打开的频道 ID
openPublicProfile	打开公众号资料卡,可以通过 public-id 属性设定需要打开的公众号资料卡的账号,同时在 manifest.json 中必须配置 publicIdList
shareMessageToFriend	在自定义开放数据域组件中向指定好友发起分享
addColorSign	添加彩签,点击后添加状态有用户提示,无回调
addGroupApp	添加群应用(只有群管理员或群主有权操作),添加后给按钮绑定@addgroupapp 事件接收回调数据
addToFavorites	收藏当前页面,点击按钮后会触发 Page.onAddToFavorites 方法
chooseAddress	选择用户收货地址,可以从@chooseaddress 回调中获取用户选择的地址信息
chooseInvoiceTitle	选择用户发票抬头,可以从@chooseinvoicetitle 回调中获取用户选择的发票抬头信息
login	登录,可以从@login 回调中确认是否登录成功
subscribe	订阅类模板消息,需要用户授权才可发送
favorite	触发用户收藏操作
watchLater	触发用户稍后再看操作
openProfile	触发打开用户主页操作

button 的点击遵循 Vue 标准的 @click 事件。button 没有 url 属性,如果要跳转页面,可以在@click 中编写代码实现,也可以在 button 外面套一层 navigator 组件(navigator 组件参见 2.5.3 小节)。如需跳转到 about 页面,可按以下几种方法实现。

```
<template>
    <view>
        <navigator url="/pages/about/about"><button type="default">通过 navigator 组件跳转到 about 页面</button></navigator>
```

```
            <button type="default" @click="goto('/pages/about/about')">通过方法跳
转到 about 页面</button>
            <button type="default" @click="navigateTo('/pages/about/about')">跳转
到 about 页面</button><!-- 这种写法只有 H5 平台支持，不跨平台，不推荐使用 -->
        </view>
    </template>
    <script>
        export default {
            methods:{
                goto(url) {
                    uni.navigateTo({
                        url:url
                    })
                }
            }
        }
    </script>
```

【实例 4-5】演示 button 的使用。

实现步骤

（1）在项目 uniappch04 中新建一个名为"button"的页面并勾选【在 pages.json 中注册】。

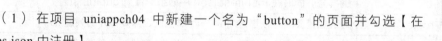
实例 4-5

（2）在 pages.json 文件中将刚创建页面的 navigationBarTitleText 设为 button。

（3）在 button.vue 中完成如下代码的编写。

```
<template>
    <view>
        <view class="uni-padding-wrap uni-common-mt">
            <button type="primary">页面主操作 Normal</button>
            <button type="primary" loading="true">页面主操作 Loading</button>
            <button type="primary" disabled="true">页面主操作 Disabled</button>
            <button type="default">页面次要操作 Normal</button>
            <button type="default" disabled="true">页面次要操作 Disabled</button>
            <button type="warn">警告类操作 Normal</button>
            <button type="warn" disabled="true">警告类操作 Disabled</button>
            <view class="button-sp-area">
                <button type="primary" plain="true">按钮</button>
                <button type="primary" disabled="true" plain="true">不可点击的
按钮</button>
                <button type="default" plain="true">按钮</button>
                <button type="default" disabled="true" plain="true">按钮</button>
                <button class="mini-btn" type="primary" size="mini">按钮</button>
                <button class="mini-btn" type="default" size="mini">按钮</button>
                <button class="mini-btn" type="warn" size="mini">按钮</button>
            </view>
        </view>
    </view>
</template>
<style>
```

```
    button{margin:20rpx 0;}
    .button-sp-area {
        margin:0 auto;
        width:60%;
    }
    .uni-padding-wrap{
        /* width:690rpx; */
        padding:0 30rpx;
    }
    .uni-common-mt{
        margin-top:30rpx;
    }
</style>
```

(4)在当前页面选择【运行】—【运行到浏览器】,即可查看运行效果,如图 4-5 所示。

图 4-5 button 演示效果

4.3.2 picker

picker(选择器)组件是从底部弹起的滚动选择器。picker 支持 5 种选择器,通过 mode 属性来设置,分别是普通选择器、多列选择器、时间选择器、日期选择器、省市区选择器,默认是普通选择器。

当 mode 属性值为 selector 或没有指定 mode 属性时,为普通选择器。普通选择器的属性说明如表 4-9 所示。

表 4-9 普通选择器的属性说明

属性	类型	默认值	说明
range	Array / Array<Object>	[]	mode 为 selector 或 multiSelector 时,range 有效

续表

属性	类型	默认值	说明
range-key	String		当 range 是 Array<Object> 类型时,通过 range-key 来指定 Object 中 key 的值作为选择器的显示内容
value	Number	0	value 的值表示选择的是 range 中的第几个元素(索引从 0 开始)
selector-type	String	auto	大屏显示时 UI 类型,支持 picker、select、auto,默认在 iPad 中以 picker 样式展示,而在 PC 中以 select 样式展示
disabled	Boolean	false	是否禁用
@change	EventHandle		value 改变时触发 change 事件,event.detail = {value: value}
@cancel	EventHandle		取消选择或点遮罩层收起 picker 时触发 cancel 事件

平台差异说明:selector-type 仅 H5 2.9.9+支持,快手小程序不支持 disabled、@cancel 属性。

当 mode 属性值为 multiSelector 时,picker 为多列选择器。多列选择器的使用非常灵活,可根据自定义的数组列数显示。多列选择器的属性说明如表 4-10 所示。

表 4-10 多列选择器的属性说明

属性	类型	默认值	说明
range	二维 Array 或二维 Array<Object>	[]	mode 为 selector 或 multiSelector 时,range 有效。二维数组的长度表示列数,数组的每项表示每列的数据,如[["a","b"], ["c","d"]]
range-key	String		当 range 是二维 Array<Object> 类型时,通过 range-key 来指定 Object 中 key 的值作为选择器显示的内容
value	Array	[]	value 的值表示选择的是 range 对应项中的第几个元素(索引从 0 开始)
@change	EventHandle		value 改变时触发 change 事件,event.detail = {value: value}
@columnchange	EventHandle		某一列的值改变时触发 columnchange 事件,event.detail = {column: column, value: value},column 的值表示改变值的是第几列(索引从 0 开始),value 的值表示变更值的索引
@cancel	EventHandle		取消选择时触发 cancel 事件(快手小程序不支持)
disabled	Boolean	false	是否禁用(快手小程序不支持)

当 mode 属性值为 time 时,picker 为时间选择器。时间选择器的属性说明如表 4-11 所示。

表 4-11 时间选择器的属性说明

属性	类型	默认值	说明
value	String		表示选中的时间，格式为"hh:mm"
start	String		表示有效时间范围的开始，字符串格式为"hh:mm"
end	String		表示有效时间范围的结束，字符串格式为"hh:mm"
@change	EventHandle		value 改变时触发 change 事件，event.detail = {value: value}
@cancel	EventHandle		取消选择时触发 cancel 事件
disabled	Boolean	false	是否禁用
value	String		表示选中的时间，格式为"hh:mm"

当 mode 属性值为 date 时，picker 为日期选择器。日期选择器的属性说明如表 4-12 所示。

表 4-12 日期选择器的属性说明

属性	类型	默认值	说明
value	String	0	表示选中的日期，格式为"YYYY-MM-DD"
start	String		表示有效日期范围的开始，字符串格式为"YYYY-MM-DD"
end	String		表示有效日期范围的结束，字符串格式为"YYYY-MM-DD"
fields	String	day	有效值为 year、month、day，表示选择器的粒度，默认值为 day，App 平台未配置此属性时使用系统 UI
@change	EventHandle		value 改变时触发 change 事件，event.detail = {value: value}
@cancel	EventHandle		取消选择时触发 cancel 事件
disabled	Boolean	false	是否禁用
value	String	0	表示选中的日期，格式为"YYYY-MM-DD"

当 mode 属性值为 region 时，picker 为省市区选择器。省市区选择器的属性说明如表 4-13 所示。

表 4-13 省市区选择器的属性说明

属性	类型	默认值	说明
value	Array	[]	表示选中的省市区，默认选中每一列的第一个值
custom-item	String		可在每一列的顶部添加一个自定义的项
@change	EventHandle		value 改变时触发 change 事件，event.detail = {value: value}
@cancel	EventHandle		取消选择时触发 cancel 事件
disabled	Boolean	false	是否禁用

【实例 4-6】演示 picker 的使用。

 实现步骤

（1）在项目 uniappch04 中新建一个名为"picker"的页面并勾选【在 pages.json 中注册】。

实例 4-6

（2）在pages.json文件中将刚创建页面的navigationBarTitleText设为picker。
（3）在picker.vue中完成如下代码的编写。

```
<template>
    <view>
        <view>地区选择器</view>
        <view>
            <view class="uni-list-cell">
                <view class="uni-list-cell-left">
                    当前选择
                </view>
                <view class="uni-list-cell-db">
                    <picker @change="bindPickerChange" :value="index" :range="array">
                        <view class="uni-input">{{array[index]}}</view>
                    </picker>
                </view>
            </view>
        </view>
        <view>时间选择器</view>
        <view>
            <view class="uni-list-cell">
                <view class="uni-list-cell-left">
                    当前选择
                </view>
                <view class="uni-list-cell-db">
                    <picker mode="time" :value="time" start="09:01" end="21:01" @change="bindTimeChange">
                        <view class="uni-input">{{time}}</view>
                    </picker>
                </view>
            </view>
        </view>
        <view>日期选择器</view>
        <view>
            <view class="uni-list-cell">
                <view class="uni-list-cell-left">
                    当前选择
                </view>
                <view class="uni-list-cell-db">
                    <picker mode="date" :value="date" :start="startDate" :end="endDate" @change="bindDateChange">
                        <view class="uni-input">{{date}}</view>
                    </picker>
                </view>
            </view>
        </view>
    </view>
</template>
<script>
    export default {
```

```
    data() {
        const currentDate = this.getDate({
            format:true
        })
        return {
            title:'picker',
            array:['中国', '美国', '巴西', '日本'],
            index:0,
            date:currentDate,
            time:'12:01'
        }
    },
    computed:{
        startDate() {
            return this.getDate('start');
        },
        endDate() {
            return this.getDate('end');
        }
    },
    methods:{
        bindPickerChange:function(e) {
            console.log('picker发送选择改变，携带值为', e.detail.value)
            this.index = e.detail.value
        },
        bindDateChange:function(e) {
            this.date = e.detail.value
        },
        bindTimeChange:function(e) {
            this.time = e.detail.value
        },
        getDate(type) {
            const date = new Date();
            let year = date.getFullYear();
            let month = date.getMonth() + 1;
            let day = date.getDate();

            if (type === 'start') {
                year = year - 60;
            } else if (type === 'end') {
                year = year + 2;
            }
            month = month > 9 ? month :'0' + month;
            day = day > 9 ? day :'0' + day;
            return '${year}-${month}-${day}';
        }
    }
}
</script>
<style>
    .uni-list-cell {
        position:relative;
        display:flex;
```

```
        flex-direction:row;
        justify-content:space-between;
        align-items:center;
    }
    .uni-list-cell-left {
        white-space:nowrap;
        font-size:28rpx;
        padding:0 30rpx;
    }
    .uni-list-cell-db{
        flex:1;
    }
    .uni-input {
        height:50rpx;
        padding:15rpx 25rpx;
        line-height:50rpx;
        font-size:28rpx;
        background:#FFF;
        flex:1;
    }
</style>
```

（4）在当前页面选择【运行】—【运行到浏览器】，即可查看运行效果，如图4-6所示。

图 4-6 picker 演示效果

4.3.3 slider

slider（滑动）组件经常在控制声音大小、屏幕亮度等场景中使用，它可以对滑动步长、最大值、最小值等进行设置，其属性说明如表4-14所示。

表 4-14 slider 的属性说明

属性	类型	默认值	说明
min	Number	0	最小值
max	Number	100	最大值
step	Number	1	步长，其值必须大于 0，并且可被（max-min）整除
disabled	Boolean	false	是否禁用
value	Number	0	当前值
activeColor	Color		滑块左侧已选择部分的线条颜色
backgroundColor	Color	#e9e9e9	滑块右侧背景线条的颜色
block-size	Number	28	滑块的大小，取值范围为 12～28
block-color	Color	#ffffff	滑块的颜色
show-value	Boolean	false	是否显示当前值
@change	EventHandle		完成一次拖动后触发的事件，event.detail = {value: value}
@changing	EventHandle		拖动过程中触发的事件，event.detail = {value: value}

【实例 4-7】演示 slider 的使用。

实例 4-7

实现步骤

（1）在项目 uniappch04 中新建一个名为"slider"的页面并勾选【在 pages.json 中注册】。

（2）在 pages.json 文件中将刚创建页面的 navigationBarTitleText 设为 slider。

（3）在 slider.vue 中完成如下代码的编写。

```
<template>
    <view>
        <view class="uni-padding-wrap uni-common-mt">
            <view class="uni-title">设置 step</view>
            <view>
                <slider value="60" @change="sliderChange" step="5" />
            </view>
            <view class="uni-title">显示当前 value</view>
            <view>
                <slider value="50" @change="sliderChange" show-value />
            </view>
            <view class="uni-title">设置最小值/最大值</view>
            <view>
                <slider value="100" @change="sliderChange" min="50" max="200" show-value />
            </view>
            <view class="uni-title">不同颜色和大小的滑块</view>
            <view>
                <slider value="50" @change="sliderChange" activeColor="#FFCC33" backgroundColor="#000000" block-color="#8A6DE9" block-size="20" />
```

```
            </view>
        </view>
    </view>
</template>
<script>
export default {
    data() {
        return {}
    },
    methods:{
        sliderChange(e) {
            console.log('value 发生变化:' + e.detail.value)
        }
    }
}
</script>
<style>
    .uni-padding-wrap{
        /* width:690rpx; */
        padding:0 30rpx;
    }
    .uni-common-mt{
        margin-top:30rpx;
    }
    .uni-title {
        font-size:30rpx;
        font-weight:500;
        padding:20rpx 0;
        line-height:1.5;
    }
</style>
```

（4）在当前页面选择【运行】—【运行到浏览器】，即可查看运行效果，如图4-7所示。

4.3.4 input

uni-app 的 input（单行输入框）组件用来输入单行文本内容，它的功能非常强大，使用方式与传统 H5 的 <input></input> 标签相似，同样支持 v-model。input 的属性说明如表 4-15 所示。

图 4-7 slider 演示效果

表 4-15 input 的属性说明

属性	类型	默认值	说明
value	String		输入框的初始内容
type	String	text	输入框的类型
text-content-type	String		文本区域的语义，根据类型自动填充
password	Boolean	false	是否是密码类型

续表

属性	类型	默认值	说明
placeholder	String		输入框为空时的占位符
placeholder-style	String		指定 placeholder 的样式
placeholder-class	String	"input-placeholder"	指定 placeholder 的样式类，注意页面或组件的 <style></style>标签中写了 scoped 时，需要在类名前写/deep/
disabled	Boolean	false	是否禁用该输入框
maxlength	Number	140	最大输入长度，设置为-1 的时候不限制最大输入长度
cursor-spacing	Number	0	指定光标与软键盘的距离，单位为 px。取输入框与底部的距离和 cursor-spacing 指定的距离的最小值作为光标与软键盘的距离
focus	Boolean	false	获取焦点
confirm-type	String	done	设置软键盘右下角按钮的文字，仅在 type="text" 时生效
confirm-hold	Boolean	false	点击软键盘右下角按钮时是否保持软键盘不收起
cursor	Number		指定聚焦时的光标位置
selection-start	Number	-1	光标起始位置，自动聚焦时有效，需与 selection-end 搭配使用
selection-end	Number	-1	光标结束位置，自动聚焦时有效，需与 selection-start 搭配使用
adjust-position	Boolean	true	软键盘弹起时，是否自动上推页面
auto-blur	Boolean	false	软键盘收起时，是否自动失去焦点
ignoreCompositionEvent	Boolean	true	是否忽略组件内对文本合成系统事件的处理。该属性为 false 时将触发 compositionstart、compositionend、compositionupdate 事件，且在文本合成期间会触发 input 事件
always-embed	Boolean	false	强制 input 处于同层状态，默认在聚焦时 input 会切换到非同层状态（仅在 iOS 下生效）
hold-keyboard	Boolean	false	聚焦时，点击页面的时候不收起软键盘
safe-password-cert-path	String		使用安全软键盘加密公钥的路径，只支持包内路径
safe-password-length	Number		使用安全软键盘输入密码的长度
safe-password-time-stamp	Number		使用安全软键盘加密时间戳
safe-password-nonce	String		使用安全软键盘加密盐值
safe-password-salt	String		使用安全软键盘计算哈希盐值，若指定 safe-password-custom-hash 则无效
safe-password-custom-hash	String		使用安全软键盘计算哈希的算法表达式，如 md5(sha1('foo' + sha256(sm3(password + 'bar'))))

续表

属性	类型	默认值	说明
random-number	Boolean	false	当 type 为 number、digit、idcard 时，数字软键盘是否随机排列
controlled	Boolean	false	是否为受控组件。此属性为 true 时，value 的值会完全受 setData 控制
always-system	Boolean	false	是否强制使用系统软键盘和 WebView 创建的 input 元素。此属性为 true 时，confirm-type、confirm-hold 可能失效
@input	EventHandle		当使用软键盘输入时，触发 input 事件，event.detail = {value}
@focus	EventHandle		输入框聚焦时触发 focus 事件，event.detail = { value, height }，height 为软键盘高度
@blur	EventHandle		输入框失去焦点时触发 blur 事件，event.detail = {value: value}
@confirm	EventHandle		点击【完成】按钮时触发 confirm 事件，event.detail = {value: value}
@keyboardheightchange	eventhandle		软键盘高度发生变化的时候触发 keyboardheightchange 事件，event.detail = {height: height, duration: duration}

平台差异说明：其中 random-number、controlled、always-system 属性仅支付宝小程序支持；@blur、@confirm 属性不被快手小程序支持；hold-keyboard、safe-password-cert-path、safe-password-length、safe-password-time-stamp、safe-password-nonce、safe-password-salt、safe-password-custom-hash、@keyboardheightchange 仅被微信小程序支持。更多细节请参考 uni-app 官网。

【**实例 4-8**】演示 input 的使用。

实例 4-8

⚙ **实现步骤**

（1）在项目 uniappch04 中新建一个名为"input"的页面并勾选【在 pages.json 中注册】。

（2）在 pages.json 文件中将刚创建页面的 navigationBarTitleText 设为 input。

（3）在 input.vue 中完成如下代码的编写。

```
<template>
    <view>
        <view class="uni-common-mt">
            <view class="uni-form-item uni-column">
                <view class="title">可自动聚焦的 input</view>
                <input class="uni-input" focus placeholder="自动获得焦点" />
            </view>
            <view class="uni-form-item uni-column">
                <view class="title">软键盘右下角按钮显示为搜索</view>
                <input class="uni-input" confirm-type="search" placeholder="软键盘右下角按钮显示为搜索" />
            </view>
```

```html
        <view class="uni-form-item uni-column">
            <view class="title">控制最大输入长度的input</view>
            <input class="uni-input" maxlength="10" placeholder="最大输入长度为10" />
        </view>
        <view class="uni-form-item uni-column">
            <view class="title">实时获取输入值:{{inputValue}}</view>
            <input class="uni-input" @input="onKeyInput" placeholder="输入同步到view中" />
        </view>
        <view class="uni-form-item uni-column">
            <view class="title">控制输入的input</view>
            <input class="uni-input" @input="replaceInput" v-model="changeValue" placeholder="连续的两个1会变成2" />
        </view>
        <!-- #ifndef MP-BAIDU -->
        <view class="uni-form-item uni-column">
            <view class="title">控制软键盘的input</view>
            <input class="uni-input" ref="input1" @input="hideKeyboard" placeholder="输入123自动收起软键盘" />
        </view>
        <!-- #endif -->
        <view class="uni-form-item uni-column">
            <view class="title">输入数字的input</view>
            <input class="uni-input" type="number" placeholder="这是一个数字输入框" />
        </view>
        <view class="uni-form-item uni-column">
            <view class="title">输入密码的input</view>
            <input class="uni-input" password type="text" placeholder="这是一个密码输入框" />
        </view>
        <view class="uni-form-item uni-column">
            <view class="title">带小数点的input</view>
            <input class="uni-input" type="digit" placeholder="带小数点的数字输入框" />
        </view>
        <view class="uni-form-item uni-column">
            <view class="title">输入身份证的input</view>
            <input class="uni-input" type="idcard" placeholder="身份证输入框" />
        </view>
        <view class="uni-form-item uni-column">
            <view class="title">控制占位符颜色的input</view>
            <input class="uni-input" placeholder-style="color:#F76260" placeholder="占位符是红色的" />
        </view>
    </view>
```

```
        </view>
    </template>
    <script>
    export default {
        data() {
            return {
                title:'input',
                focus:false,
                inputValue:'',
                changeValue:''
            }
        },
        methods:{
            onKeyInput:function(event) {
                this.inputValue = event.target.value
            },
            replaceInput:function(event) {
                var value = event.target.value;
                if (value === '11') {
                    this.changeValue = '2';
                }
            },
            hideKeyboard:function(event) {
                if (event.target.value === '123') {
                    uni.hideKeyboard();
                }
            }
        }
    }
    </script>
    <style>
        .uni-common-mt{
            margin-top:30rpx;
        }
        .uni-form-item{
            display:flex;
            width:100%;
            padding:10rpx 0;
        }
        .uni-column {
            flex-direction:column;
        }
        .uni-input {
            height:50rpx;
            padding:15rpx 25rpx;
            line-height:50rpx;
            font-size:28rpx;
            background:#FFF;
            flex:1;
        }
    </style>
```

（4）在当前页面选择【运行】—【运行到浏览器】，即可查看运行效果，如图4-8所示。

图 4-8　input 演示效果

4.3.5　radio 和 checkbox

uni-app 的 radio（单选按钮）组件与传统 H5 的<radio></radio>标签的使用方式不太一样，需要使用 radio-group 组件包裹。

radio-group 由多个 radio 组成。通过把多个 radio 包裹在一个 radio-group 中，实现这些 radio 的单选。radio 的属性说明如表 4-16 所示。

表 4-16　radio 的属性说明

属性	类型	默认值	说明
@change	EventHandle		radio-group 中的选中项发生变化时触发 change 事件，event.detail = {value: 选中的 radio 的 value}
value	String		radio 标识。当某个 radio 被选中时 radio-group 的 change 事件会携带 radio 的 value
checked	Boolean	false	当前是否被选中
disabled	Boolean	false	是否禁用
color	Color		radio 的颜色，同 CSS 的 color

uni-app 的 checkbox（复选框）组件与传统 H5 的<checkbox></checkbox>标签的使用方式不太一样，需要使用 checkbox-group 组件包裹。

check-group 由多个 checkbox 组成。checkbox 的属性说明如表 4-17 所示。

表 4-17 checkbox 的属性说明

属性	类型	默认值	说明
@change	EventHandle		checkbox-group 中选中项发生改变时触发 change 事件，detail = {value:[选中的 checkbox 的 value 的数组]}
value	String		checkbox 的标识，选中某个 checkbox 时触发 checkbox-group 的 change 事件，并携带 checkbox 的 value
disabled	Boolean	false	是否禁用
checked	Boolean	false	当前是否被选中，可用来设置默认选中
color	Color		checkbox 的颜色，同 CSS 的 color

radio 和 checkbox 往往与 label 组件一起使用。label 组件常用来改进表单组件的可用性。它有两种使用方式，一是使用 label 组件的 for 属性关联对应表单组件的 id，二是将表单组件放在<label></label>中，点击时，就会触发对应的表单组件。

【实例 4-9】演示 radio 和 checkbox 的使用。

实现步骤

（1）在项目 uniappch04 中新建一个名为"radio&checkbox"的页面并勾选【在 pages.json 中注册】。

实例 4-9

（2）在 pages.json 文件中将刚创建页面的 navigationBarTitleText 设为 radio&checkbox。

（3）在 radio&checkbox.vue 中完成如下代码的编写。

```
<template>
    <view>
        <view class="uni-common-mt">
            <view class="uni-form-item uni-column">
                <view class="title">表单组件在&lt;label&gt;&lt;/label&gt;内</view>
                <checkbox-group class="uni-list" @change="checkboxChange">
                    <label class="uni-list-cell uni-list-cell-pd" v-for="item in checkboxItems" :key="item.name">
                        <view>
                            <checkbox :value="item.name" :checked="item.checked"></checkbox>
                        </view>
                        <view>{{item.value}}</view>
                    </label>
                </checkbox-group>
            </view>

            <view class="uni-form-item uni-column">
                <view class="title">label 组件用 for 属性标识表单组件</view>
                <radio-group class="uni-list" @change="radioChange">
                    <label class="uni-list-cell uni-list-cell-pd" v-for="(item,index) in radioItems" :key="index">
                        <view>
                            <radio :id="item.name":value="item.name" :checked="item.checked"></radio>
                        </view>
```

```html
                    <view>
                        <label class="label-2-text" :for="item.name">
                            <text>{{item.value}}</text>
                        </label>
                    </view>
                </label>
            </radio-group>
        </view>
    </view>
</view>
</template>
<script>
export default {
    data() {
        return {
            checkboxItems:[ {
                    name:'USA',
                    value:'美国'
                },  {
                    name:'CHN',
                    value:'中国',
                    checked:'true'
                }  ],
            radioItems:[ {
                    name:'USA',
                    value:'美国'
                },  {
                    name:'CHN',
                    value:'中国',
                    checked:'true'
                }
            ],
            hidden:false
        }
    },
    methods:{
        checkboxChange:function(e) {
            console.log(e)
            var checked = e.target.value
            var changed = {}
            for (var i = 0; i < this.checkboxItems.length; i++) {
                if (checked.indexOf(this.checkboxItems[i].name) !== -1) {
                    changed['checkboxItems[' + i + '].checked'] = true
                } else {
                    changed['checkboxItems[' + i + '].checked'] = false
                }
            }
        },
        radioChange:function(e) {
            var checked = e.target.value
            var changed = {}
            for (var i = 0; i < this.radioItems.length; i++) {
                if (checked.indexOf(this.radioItems[i].name) !== -1) {
```

```
                    changed['radioItems[' + i + '].checked'] = true
                } else {
                    changed['radioItems[' + i + '].checked'] = false
                }
            }
        }
    }
}
</script>
<style>
    .uni-common-mt{
        margin-top:30rpx;
    }
    .uni-form-item{
        display:flex;
        width:100%;
        padding:10rpx 0;
    }
    .uni-form-item .title{
        padding:10rpx 25rpx;
    }
    .uni-column {
        flex-direction:column;
    }
    .uni-list {
        background-color:#FFFFFF;
        position:relative;
        width:100%;
        display:flex;
        flex-direction:column;
    }
    .uni-list-cell {
        position:relative;
        display:flex;
        flex-direction:row;
        justify-content:space-between;
        align-items:center;
    }
    .uni-list-cell-pd {
        padding:22rpx 30rpx;
    }
</style>
```

（4）在当前页面选择【运行】—【运行到浏览器】，即可查看运行效果，如图4-9所示。

4.3.6 switch

switch（开关选择器）组件在实际开发中经常使用，如应用于是否显示密码、是否打包、是否通知、是否使用余额等场景。switch 的属性说明如表4-18所示。

图4-9 radio 和 checkbox 演示效果

表 4-18　switch 的属性说明

属性	类型	默认值	说明
checked	Boolean	false	是否选中
disabled	Boolean	false	是否禁用
type	String	switch	样式，有效值：switch、checkbox
color	Color		颜色，同 CSS 的 color
@change	EventHandle		checked 改变时触发 change 事件，event.detail={ value:checked}

平台差异说明：字节跳动、飞书小程序不支持 disabled。

switch 的默认颜色在不同平台上不一样，如在微信小程序中是绿色的，在字节跳动小程序中是红色的，在其他平台是蓝色的。更改颜色使用 color 属性。如需调节 switch 的大小，可通过 CSS 的 scale 方法，如将组件缩小到原来的 70%：style="transform:scale(0.7)"。

【实例 4-10】演示 switch 的使用。

实例 4-10

※ **实现步骤**

（1）在项目 uniappch04 中新建一个名为"switch"的页面并勾选【在 pages.json 中注册】。

（2）在 pages.json 文件中将刚创建页面的 navigationBarTitleText 设为 switch。

（3）在 switch.vue 中完成如下代码的编写。

```
<template>
    <view>
        <view class="uni-padding-wrap uni-common-mt">
            <view class="uni-title">默认样式</view>
            <view>
                <switch checked @change="switch1Change" />
                <switch @change="switch2Change" />
            </view>
            <view class="uni-title">不同颜色和尺寸的switch</view>
            <view>
                <switch checked color="#FFCC33" style="transform:scale(0.7)"/>
                <switch color="#FFCC33" style="transform:scale(0.7)"/>
            </view>
        </view>
    </view>
</template>
<script>
export default {
    data() {
        return {}
    },
    methods:{
        switch1Change:function (e) {
            console.log('switch1 发生 change 事件，携带值为', e.detail.value)
        },
        switch2Change:function (e) {
```

```
                console.log('switch2 发生 change 事件,携带值为', e.detail.value)
            }
        }
    }
</script>
<style>
    .uni-padding-wrap{
        /* width:690rpx; */
        padding:0 30rpx;
    }
    .uni-common-mt{
        margin-top:30rpx;
    }
    .uni-title {
        font-size:30rpx;
        font-weight:500;
        padding:20rpx 0;
        line-height:1.5;
    }
</style>
```

（4）在当前页面选择【运行】—【运行到浏览器】，即可查看运行效果，如图 4-10 所示。

4.3.7 textarea

textarea（多行输入框）组件是与 input（单行输入框）组件对应的组件，它的使用方法与传统的 H5 的 \<textarea>\</textarea>标签的使用方法类似，其功能非常强大，在实际开发中经常应用于填写备注、个人介绍等场景。textarea 的属性说明如表 4-19 所示。

图 4-10 switch 演示效果

表 4-19 textarea 的属性说明

属性名	类型	默认值	说明
value	String		输入框的内容
placeholder	String		输入框为空时的占位符
placeholder-style	String		指定 placeholder 的样式
placeholder-class	String	textarea-placeholder	指定 placeholder 的样式类，注意页面或组件的\<style>\</style>标签中写了 scoped 时，需要在类名前写/deep/
disabled	Boolean	false	是否禁用
maxlength	Number	140	最大输入长度，设置为-1 的时候不限制最大输入长度
focus	Boolean	false	获取焦点
auto-focus	Boolean	false	自动聚焦，收起软键盘

续表

属性名	类型	默认值	说明
auto-height	Boolean	false	是否自动增高，设置 auto-height 时，style.height 不生效
fixed	Boolean	false	如果输入框是一个在 position:fixed 的区域，需要显式指定属性 fixed 为 true
cursor-spacing	Number	0	指定光标与软键盘的距离，单位为 px。取输入框与底部的距离和 cursor-spacing 指定的距离的最小值作为光标与软键盘的距离
cursor	Number		指定聚焦时的光标位置
confirm-type	String	done	设置软键盘右下角按钮的文字
confirm-hold	Boolean	false	点击软键盘右下角按钮时是否保持软键盘不收起
show-confirm-bar	Boolean	true	是否显示软键盘上方带有【完成】按钮的那一栏
selection-start	Number	-1	光标起始位置，自动聚焦时有效，需与 selection-end 搭配使用
selection-end	Number	-1	光标结束位置，自动聚焦时有效，需与 selection-start 搭配使用
adjust-position	Boolean	true	软键盘弹起时，是否自动上推页面
disable-default-padding	Boolean	false	是否去掉 iOS 下的默认内边距
hold-keyboard	Boolean	false	聚焦时，点击页面的时候不收起软键盘
auto-blur	Boolean	false	软键盘收起时，是否自动失去焦点
ignoreCompositionEvent	Boolean	true	是否忽略组件内对文本合成系统事件的处理。此属性为 false 时将触发 compositionstart、compositionend、compositionupdate 事件，且在文本合成期间会触发 input 事件
@focus	EventHandle		输入框聚焦时触发 focus 事件，event.detail = { value, height }，height 为软键盘高度
@blur	EventHandle		输入框失去焦点时触发 blur 事件，event.detail = {value, cursor}
@linechange	EventHandle		输入框行数变化时触发 linechange 事件，event.detail = {height: 0, heightRpx: 0, lineCount: 0}
@input	EventHandle		当使用软键盘输入时，触发 input 事件，event.detail = {value, cursor}，@input 处理函数的返回值并不会反映到 textarea 上
@confirm	EventHandle		点击【完成】按钮时，触发 confirm 事件，event.detail = {value: value}
@keyboardheightchange	EventHandle		软键盘高度发生变化的时候触发 keyboardheightchange 事件，event.detail = {height: height, duration: duration}

平台差异说明：快手小程序不支持@blur、@input。更多的差异细节请参见 uni-app 官网。

【实例 4-11】演示 textarea 的使用。

实现步骤

实例 4-11

（1）在项目 uniappch04 中新建一个名为"textarea"的页面并勾选【在 pages.json 中注册】。

（2）在 pages.json 文件中将刚创建页面的 navigationBarTitleText 设为 textarea。

（3）在 textarea.vue 中完成如下代码的编写。

```
<template>
    <view>
        <view class="uni-title uni-common-pl">输入框高度自适应，不会出现滚动条</view>
        <view class="uni-textarea">
            <textarea @blur="bindTextAreaBlur" auto-height />
        </view>
        <view class="uni-title uni-common-pl">占位符是红色的</view>
        <view class="uni-textarea">
            <textarea placeholder-style="color:#F76260" placeholder="占位符是红色的"/>
        </view>
    </view>
</template>
<script>
export default {
    data() {
        return {}
    },
    methods:{
        bindTextAreaBlur:function (e) {
            console.log(e.detail.value)
        }
    }
}
</script>
<style>
    .uni-title {
        font-size:30rpx;
        font-weight:500;
        padding:20rpx 0;
        line-height:1.5;
    }
    .uni-common-pl{
        padding-left:30rpx;
    }
    .uni-textarea{
        width:100%;
        background:#FFF;
    }
    .uni-textarea textarea{
        width:96%;
        padding:18rpx 2%;
        line-height:1.6;
        font-size:28rpx;
```

```
            height:180rpx;
        }
</style>
```

（4）在当前页面选择【运行】—【运行到浏览器】，即可查看运行效果，如图 4-11 所示。

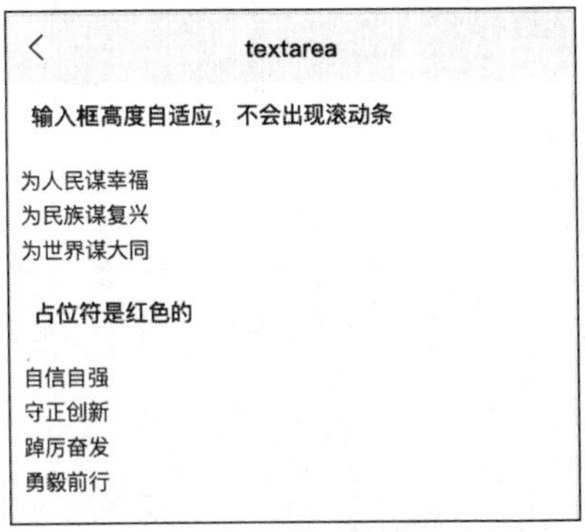

4.3.8 form

form 组件将其中的 switch、input、slider、radio、picker 等表单组件的用户输入内容进行提交。当点击 <form></from> 中的提交按钮（formType 属性为"submit"的 button 组件）时，会将 <form></from> 中各个表单组件的 value 属性的值进行提交。注意：<form></from> 中各个表单组件需设置 name 属性，提交表单时，name 属性的值作为 key。form 的属性说明如表 4-20 所示。

图 4-11 textarea 演示效果

表 4-20 form 的属性说明

属性	类型	说明
report-submit	Boolean	是否返回 formId 用于发送模板消息
report-submit-timeout	Number	等待一段时间（以 ms 为单位）以确认 formId 是否生效。如果未指定这个参数，formId 有很小的概率是无效的（如遇到网络失败的情况）。指定这个参数将可以检测 formId 是否有效，以这个参数的值作为检测的超时时间。如果失败，将返回以 requestFormId:fail 开头的 formId
@submit	EventHandle	携带 form 中的数据触发 submit 事件，event.detail = {value: {'name': 'value'}, formId: ''}，report-submit 为 true 时才会返回 formId
@reset	EventHandle	表单重置时会触发 reset 事件

平台差异说明：微信、支付宝小程序支持 report-submit 属性；微信小程序 2.6.2 支持 report-submit-timeout 属性，其他平台不支持。

【实例 4-12】演示 form 的使用。

实现步骤

（1）在项目 uniappch04 中新建一个名为"form"的页面并勾选【在 pages.json 中注册】。

（2）在 pages.json 文件中将刚创建页面的 navigationBarTitleText 设为 form。

（3）在 form.vue 中完成如下代码的编写。

```
<template>
    <view>
        <view class="uni-padding-wrap uni-common-mt">
            <form @submit="formSubmit" @reset="formReset">
```

```
            <view class="uni-form-item uni-column">
                <view class="title">姓名</view>
                <input class="uni-input" name="nickname" placeholder="请输入姓名" />
            </view>
            <view class="uni-form-item uni-column">
                <view class="title">性别</view>
                <radio-group name="gender">
                    <label>
                        <radio value="男" /><text>男</text>
                    </label>
                    <label>
                        <radio value="女" /><text>女</text>
                    </label>
                </radio-group>
            </view>
            <view class="uni-form-item uni-column">
                <view class="title">爱好</view>
                <checkbox-group name="loves">
                    <label>
                        <checkbox value="读书" /><text>读书</text>
                    </label>
                    <label>
                        <checkbox value="写字" /><text>写字</text>
                    </label>
                </checkbox-group>
            </view>
            <view class="uni-form-item uni-column">
                <view class="title">年龄</view>
                <slider value="20" name="age" show-value></slider>
            </view>
            <view class="uni-form-item uni-column">
                <view class="title">保留选项</view>
                <view>
                    <switch name="switch" />
                </view>
            </view>
            <view class="uni-btn-v">
                <button form-type="submit">Submit</button>
                <button type="default" form-type="reset">Reset</button>
            </view>
        </form>
    </view>
</view>
</template>
<script>
    export default {
        data() {
            return {
            }
```

```
            },
            methods:{
                formSubmit:function(e) {
                    console.log('form发生了submit事件,携带数据为:' + JSON.stringify(e.detail.value))
                    var formdata = e.detail.value
                    uni.showModal({
                        content:'表单数据内容:' + JSON.stringify(formdata),
                        showCancel:false
                    });
                },
                formReset:function(e) {
                    console.log('清空数据')
                }
            }
        }
</script>
<style>
    .uni-padding-wrap{
        padding:0 30rpx;
    }
    .uni-common-mt{
        margin-top:30rpx;
    }
    radio-group, checkbox-group{
        width:100%;
    }
    radio-group label, checkbox-group label{
        padding-right:20rpx;
    }
    form {
        width:100%;
    }

    .uni-form-item {
        display:flex;
        width:100%;
        padding:10rpx 0;
    }
    .uni-form-item .title {
        padding:10rpx 25rpx;
    }
    .uni-column {
        flex-direction:column;
    }
    .uni-btn-v {
        padding:10rpx 0;
    }
    .uni-btn-v button {
        margin:20rpx 0;
    }
</style>
```

（4）在当前页面选择【运行】—【运行到浏览器】，即可查看运行效果，如图 4-12 所示。

（a）表单提交前的演示效果　　　　（b）表单提交后的演示效果

图 4-12　form 演示效果

4.4　媒体组件

uni-app 为开发者提供了一系列媒体组件，其中最常用的是 camera 和 video。

4.4.1　camera

camera 组件是页面内嵌的区域相机组件，注意这不是点击后全屏打开的相机组件。camera 的支持平台有微信、百度、QQ、快手、京东小程序和快应用。camera 的属性说明如表 4-21 所示。

表 4-21　camera 的属性说明

属性	类型	默认值	说明
mode	String	normal	有效值为 normal、scanCode
resolution	String	medium	分辨率，不支持动态修改
device-position	String	back	前置或后置摄像头，有效值为 front、back
flash	String	auto	闪光灯，有效值为 auto、on、off
frame-size	String	medium	指定期望的相机帧尺寸
@stop	EventHandle		在摄像头非正常终止时触发 stop 事件，如退出后台等情况
@error	EventHandle		用户不允许使用摄像头时触发 error 事件
@initdone	EventHandle		相机初始化完成时触发 initdone 事件，e.detail = {maxZoom}
@scancode	EventHandle		在扫码成功时触发 scancode 事件，仅在 mode="scancode" 时生效

平台差异说明：resolution、@initdone、@scancode 仅微信小程序支持；快手小程序不支持@stop、@error；微信小程序 2.10.0、快应用支持 frame-size 属性，其他平台不支持。

camera 是由客户端创建的原生组件，它的层级是最高的，不能通过 z-index 控制层级，可使用 cover-view、cover-image 覆盖在上面。请勿在 scroll-view、swiper、picker-view、movable-view 中使用 camera。同一页面中只能插入一个 camera。

camera 的使用实例如下，需要注意的是 uni.createCameraContext 不支持 H5 及 App，需在小程序环境下使用。

```
<template>
    <view>
        <camera device-position="back" flash="off" @error="error" style="width:100%; height:300px;"></camera>
        <button type="primary" @click="takePhoto">拍照</button>
        <view>预览</view>
        <image mode="widthFix" :src="src"></image>
    </view>
</template>
<script>
    export default {
        data() {
            return {
                src:""
            }
        },
        methods:{
            takePhoto() {
                const ctx = uni.createCameraContext();
                ctx.takePhoto({
                    quality:'high',
                    success:(res) => {
                        this.src = res.tempImagePath
                    }
                });
            },
            error(e) {
                console.log(e.detail);
            }
        }
    }
</script>
```

4.4.2 video

video（视频播放）组件在 H5 平台支持 MP4、WEBM 和 OGG 格式的视频。如果在 H5 平台自行开发播放器或使用第三方视频播放器插件，系统可以自动判断环境兼容性，以决定使用传统 H5 的<video></video>标签还是 uni-app 的 video。App 平台支持本地视频（MP4/FLV）、网络视频（MP4/FLV/M3U8）及流媒体文件（RTMP/HLS/RTSP），video 的属性说明如表 4-22 所示。

表 4-22 video 的属性说明

属性	类型	默认值	说明
src	String		要播放视频的地址
autoplay	Boolean	false	是否自动播放
loop	Boolean	false	是否循环播放
muted	Boolean	false	是否静音播放
initial-time	Number		指定视频初始播放位置，单位为秒（s）
duration	Number		指定视频时长，单位为 s
controls	Boolean	true	是否显示默认播放控件（显示【播放】和【暂停】按钮、播放进度、时长）
danmu-list	Object Array		弹幕列表
danmu-btn	Boolean	false	是否显示弹幕相关按钮，只在初始化时有效，不能动态变更
enable-danmu	Boolean	false	是否显示弹幕，只在初始化时有效，不能动态变更
page-gesture	Boolean	false	在非全屏模式下，是否开启亮度与音量调节手势
direction	Number		设置全屏时视频的方向，若不指定则根据视频宽高比自动判断。有效值为 0（正常竖向）、90（屏幕逆时针 90°）、-90（屏幕顺时针 90°）
show-progress	Boolean	true	若不设置，当 video 组件宽度大于 240px 时才会显示进度条
show-fullscreen-btn	Boolean	true	是否显示【全屏】按钮
show-play-btn	Boolean	true	是否显示视频底部控制栏的【播放】按钮
show-center-play-btn	Boolean	true	是否显示视频中间的【播放】按钮
show-loading	Boolean	true	是否显示 loading 控件
enable-progress-gesture	Boolean	true	是否开启控制进度的手势
object-fit	String	contain	当视频大小与 video 大小不一致时，视频的表现形式。有效值为 contain（包含）、fill（填充）、cover（覆盖）
poster	String		视频封面的图片网络地址，如果 controls 属性值为 false，则 poster 无效
show-mute-btn	Boolean	false	是否显示【静音】按钮
title	String		视频的标题，全屏模式时在顶部展示
play-btn-position	String	bottom	【播放】按钮的位置
mobilenet-hint-type	Number	1	移动网络提醒样式：0 表示不提醒，1 表示提醒，默认值为 1
enable-play-gesture	Boolean	false	是否开启播放手势，即双击切换播放/暂停

续表

属性	类型	默认值	说明
auto-pause-if-navigate	Boolean	true	当跳转到其他小程序页面时,是否自动暂停当前页面的视频
auto-pause-if-open-native	Boolean	true	当跳转到其他微信原生页面时,是否自动暂停当前页面的视频
vslide-gesture	Boolean	false	在非全屏模式下,是否开启亮度与音量调节手势(同 page-gesture)
vslide-gesture-in-fullscreen	Boolean	true	在全屏模式下,是否开启亮度与音量调节手势
ad-unit-id	String		视频前贴广告单元 ID,更多详情可参考开放能力视频前贴广告
poster-for-crawler	String		用于给搜索等场景作为视频封面,建议使用无播放 icon 的视频封面,只支持网络地址
codec	String	hardware	解码器,有效值为 hardware(硬解码,可以增加解码算力,提高视频清晰度。少部分老旧硬件可能存在兼容性问题)、software(ffmpeg 软解码)
http-cache	Boolean	true	是否对 HTTP、HTTPS 视频源开启本地缓存策略。开启了缓存策略的视频源,在播放时会在本地保存缓存文件,如果本地缓存池已超过 100MB,在进行缓存前会清空之前的缓存(不适用于 M3U8 等流媒体格式)
header	Object		HTTP 请求 Header
@play	EventHandle		当开始/继续播放时触发 play 事件
@pause	EventHandle		当暂停播放时触发 pause 事件
@ended	EventHandle		当播放到末尾时触发 ended 事件
@timeupdate	EventHandle		播放进度变化时触发 timeupdate 事件,event.detail = {currentTime, duration}。触发频率为 250ms 一次
@fullscreenchange	EventHandle		当视频进入和退出全屏模式时触发 fullscreenchange 事件,event.detail = {fullScreen, direction},direction 取值为 vertical 或 horizontal
@waiting	EventHandle		视频缓冲时触发 waiting 事件
@error	EventHandle		视频播放出错时触发 error 事件
@progress	EventHandle		加载进度变化时触发 progress 事件,只支持一段视频的加载。event.detail = {buffered},buffered 为百分比
@loadeddata	EventHandle		视频资源开始加载时触发 loadeddata 事件
@loadstart	EventHandle		开始加载数据时触发 loadstart 事件
@seeked	EventHandle		拖动进度条结束时触发 seeked 事件
@seeking	EventHandle		正在拖动进度条时触发 seeking 事件
@loadedmetadata	EventHandle		视频元数据加载完成时触发 loadedmetadata 事件。event.detail = {width, height, duration}

续表

属性	类型	默认值	说明
@fullscreenclick	EventHandle		视频全屏播放时触发 fullscreenclick 事件。event.detail = { screenX:"Number 类型，点击点相对于屏幕左侧边缘的 x 轴坐标", screenY:"Number 类型，点击点相对于屏幕顶部边缘的 y 轴坐标", screenWidth:"Number 类型,屏幕总宽度", screenHeight:"Number 类型,屏幕总高度"}
@controlstoggle	EventHandle		切换播放控件显示隐藏时触发 controlstoggle 事件。event.detail = {show}

平台差异说明：video 属性较多，各平台支持情况较为复杂，具体细节请读者参考官方文档。

【实例 4-13】演示 video 的使用。

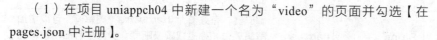 实现步骤

实例 4-13

（1）在项目 uniappch04 中新建一个名为"video"的页面并勾选【在 pages.json 中注册】。

（2）在 pages.json 文件中将刚创建页面的 navigationBarTitleText 设为 video。

（3）在 video.vue 中完成如下代码的编写。

```
<template>
    <view>
        <view class="uni-padding-wrap uni-common-mt">
            <view>
                <video id="myVideo" src="https://qiniu-web-assets.dcloud.net.cn/unidoc/zh/2minute-demo.mp4"
                    @error="videoErrorCallback" :danmu-list="danmuList" enable-danmu danmu-btn controls></video>
            </view>
        </view>
    </view>
</template>
<script>
export default {
    data() {
        return {
        }
    },
    methods:{
    }
}
</script>
<style>
    video {
        width:100%;
    }
    .uni-padding-wrap{
        /* width:690rpx; */
        padding:0 30rpx;
    }
    .uni-common-mt{
```

```
            margin-top:30rpx;
        }
</style>
```

（4）在当前页面选择【运行】—【运行到浏览器】，即可查看运行效果，如图 4-13 所示。

图 4-13　video 演示效果

4.5　地图组件

uni-app 提供了 map（地图）组件，map 用于展示地图，而定位 API 只用于获取坐标，请勿混淆两者。map 的属性过于复杂，本节将通过一个实例来演示 map 的初步使用，有需要的读者可到 uni-app 官网查阅 map 组件的相关属性配置。

【实例 4-14】演示 map 的使用。

实现步骤

（1）在项目 uniappch04 中新建一个名为"map"的页面并勾选【在 pages.json 中注册】。
（2）在 pages.json 文件中将刚创建页面的 navigationBarTitleText 设为 map。
（3）在 map.vue 中完成如下代码的编写。

```
<template>
    <view>
        <map :latitude="latitude" :longitude="longitude" :markers="covers">
        </map>
    </view>
</template>
<script>
    export default {
        data() {
            return {
                latitude:39.909,
                longitude:116.39742,
                covers:[{
                    latitude:39.909,
                    longitude:116.39742,
                    iconPath:'../../static/location.png'
                }]
            }
        },
```

```
        methods:{
        }
    }
</script>
<style>
    map {
        width:100%;
        height:600rpx;
    }
</style>
```

以上代码中 map 的 latitude 属性表示中心经度，longitude 属性表示中心纬度，markers 属性表示标记点，可以设置标记的经度、纬度及图标等。

（4）在当前页面选择【运行】—【运行到浏览器】，即可查看运行效果。

本实例中使用的是腾讯地图，需要注意的是，在使用 map 时需要在 manifest.json 文件中配置对应地图应用的 key，如图 4-14 所示。

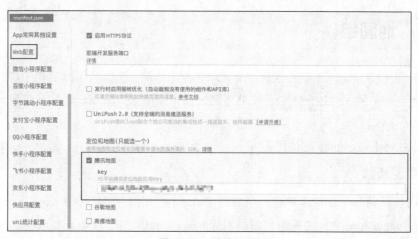

图 4-14　地图配置

4.6　案例一：典型注册页

注册是 Web、小程序等平台上多种应用的常用功能，本节将制作一个典型的注册页。页面效果如图 4-15 所示。

实现步骤

（1）新建项目。在新项目的 pages.json 中设置首页的 navigationBarTitleText 属性值为"注册"，代码如下。

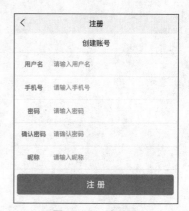

图 4-15　注册页

案例一：典型注册页

```
"pages":[    {
        "path":"pages/index/index",
        "style":{
            "navigationBarTitleText":"注册"
        }
    }    ]
```

（2）编写 index.vue 文件的<template></template>中的代码，完成元素构建，代码如下。

```
<view>
    <form bindsubmit="formSubmit" bindreset="formReset">
        <view class="loginTitle">        创建账号        </view>
        <view class="hr"></view>
        <view class="accountType">
            <view class="account">
                <view class="ac">用户名</view>
                <view class="ipt"><input name="loginName" type="text" placeholder=
"请输入用户名" class="placeholder-style" /></view>
            </view>
            <view class="hr"></view>
            <view class="account">
                <view class="ac">手机号</view>
                <view class="ipt"><input name="mobile" type="text" placeholder=
"请输入手机号" class="placeholder-style" />   </view>
            </view>
            <view class="hr"></view>
            <view class="account">
                <view class="ac">密码</view>
                <view class="ipt"><input name="loginPassword" type="text"
password placeholder="请输入密码"    class="placeholder-style" /></view>
            </view>
            <view class="hr"></view>
            <view class="account">
                <view class="ac">确认密码</view>
                <view class="ipt"><input name="confirmPassword" type="text"
password placeholder="请确认密码"    class="placeholder-style" /></view>
            </view>
            <view class="hr"></view>
            <view class="account">
                <view class="ac">昵称</view>
                <view class="ipt"><input name="nickName" type="text" placeholder=
"请输入昵称"  class="placeholder-style" /></view>
            </view>
            <view class="hr"></view>
            <view class="login">
                <button form-type="submit">注册</button>
                <view class="tip"></view>
            </view>
        </view>
    </form>
</view>
```

（3）编写 index.vue 文件的<style></style>中的代码，完成样式构建，代码如下。

```css
.content {
    height:600px;
}
.loginTitle {
    margin:10px;
    text-align:center;
}
.select {
    font-size:12px;
    color:red;
    width:50%;
    text-align:center;
    height:45px;
    line-height:45px;
    border-bottom:5rpx solid red;
}
.default {
    font-size:12px;
    margin:0 auto;
    padding:15px;
}
.hr {
    border:1px solid #cccccc;
    opacity:0.2;
}
.account {
    display:flex;
    flex-direction:row;
    align-items:center;
}
.ac {
    padding:15px;
    font-size:14px;
    font-weight:bold;
    color:#666666;
    width:60px;
    text-align:center;
}
.ipt input {
    text-align:left;
    width:200px;
}
.placeholder-style {
    font-size:14px;
    color:#cccccc;
}
.login {
    margin:0 auto;
    text-align:center;
    padding-top:10px;
}
.login button {
```

```
        width:96%;
        color:#ffffff;
        background:royalblue;
}
.tip {
        margin-top:10px;
        font-size:12px;
        color:#D53E37;
}
```

（4）在当前页面选择【运行】—【运行到浏览器】，即可查看运行效果。

4.7　案例二：典型个人中心页

个人中心页是 Web、小程序等平台上多种应用的常用功能，本节将制作一个典型的个人中心页。页面效果如图 4-16 所示。

案例二：典型
个人中心页

图 4-16　个人中心页

实现步骤

（1）新建项目。在新项目的 pages.json 中设置首页的 navigationBarTitleText 属性值为"我的"，代码如下。

```
"pages":[    {
        "path":"pages/index/index",
        "style":{
            "navigationBarTitleText":"我的"
        }
    }  ]
```

（2）编写 index.vue 文件的<template></template>中的代码，完成元素构建，代码如下。

```
    <view>
        <view class="head">
            <view class="headIcon">
                <image src="../../static/head.jpg" style="width:70px;height:70px;"></image>
            </view>
            <view class="login">     <navigator url="../login/login" hover-class="navigator-hover">{{nickName}}</navigator>     </view>
            <view class="detail"><text></text></view>
        </view>
        <view class="hr"></view>
        <view style="display:flex;flex-direction:row;">
            <view class="order">我的订单</view>
            <view class="detail2"><text></text></view>
        </view>
        <view class="line"></view>
        <view class="nav">
            <view class="nav-item" id="0">
                <view>
                    <image src="../../static/dfk.png" style="width:28px;height:25px;"></image>
                </view>
                <view>待付款</view>
            </view>
            <view class="nav-item" id="1">
                <view>
                    <image src="../../static/dsh.png" style="width:36px;height:27px;"></image>
                </view>
                <view>待收货</view>
            </view>
            <view class="nav-item" id="2">
                <view>
                    <image src="../../static/dpj.png" style="width:31px;height:28px;"></image>
                </view>
                <view>已完成</view>
            </view>
        </view>
        <view class="hr"></view>
        <view class="item">
            <view class="order">我的消息</view>
            <view class="detail2"><text></text></view>
        </view>
        <view class="line"></view>
        <view class="item">
            <view class="order">我的收藏</view>
            <view class="detail2"><text></text></view>
        </view>
        <view class="line"></view>
        <view class="item">
            <view class="order">账户余额</view>
            <view class="detail2"><text>0.00元 ></text></view>
```

```
        </view>
        <view class="line"></view>
        <view class="hr"></view>
        <view class="item">
            <view class="order">修改密码</view>
            <view class="detail2"><text></text></view>
        </view>
        <view class="line"></view>
        <view class="item">
            <view class="order">意见反馈</view>
            <view class="detail2"><text></text></view>
        </view>
        <view class="line"></view>
        <view class="item">
            <view class="order">清除缓存</view>
            <view class="detail2"><text> </text></view>
        </view>
        <view class="line"></view>
        <view class="hr"></view>
        <view class="line"></view>
        <view class="item">
            <view class="order">知识扩展</view>
            <view class="detail2"><text> </text></view>
        </view>
        <view class="hr"></view>
</view>
```

（3）编写 index.vue 文件的<style></style>中的代码，完成样式构建，代码如下。

```
.head {
        width:100%;
        height:90px;
        background-color:royalblue;
        display:flex;
        flex-direction:row;
    }
    .headIcon {
        margin:10px;
    }
    .headIcon image {
        border-radius:50%;
    }
    .login {
        color:#ffffff;
        font-size:15px;
        font-weight:bold;
        position:absolute;
        left:100px;
        margin-top:30px;
    }
    .detail {
        color:#ffffff;
        font-size:15px;
        position:absolute;
```

```css
        right:10px;
        margin-top:30px;
    }
    .nav {
        display:flex;
        flex-direction:row;
        padding-top:10px;
        padding-bottom:10px;
    }
    .nav-item {
        width:25%;
        font-size:13px;
        text-align:center;
        margin:0 auto;
    }
    .hr {
        width:100%;
        height:15px;
        background-color:#f5f5f5;
    }
    .order {
        padding-top:15px;
        padding-left:15px;
        padding-bottom:15px;
        font-size:15px;
    }
    .detail2 {
        font-size:15px;
        position:absolute;
        right:10px;
        margin-top:15px;
        color:#888888;
    }
    .line {
        height:1px;
        width:100%;
        background-color:#666666;
        opacity:0.2;
    }
    .item {
        display:flex;
        flex-direction:row;
    }
```

（4）在当前页面选择【运行】—【运行到浏览器】，即可查看运行效果。

本章小结

本章主要介绍了 uni-app 的常用组件，包括容器组件、基础组件、表单组件、媒体组件、地图组件等。这些组件的使用频率很高，读者应多加练习并记住这些组件的用法。

项目实战

使用本章及本章之前所学习的知识完成一个公司首页的制作,页面效果如图 4-17 所示。

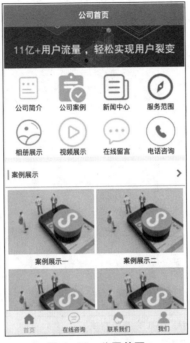

图 4-17 公司首页

拓展实训项目

主动变革、创新学习,永远是掌握时代主动的关键。面对快速变化的世界,学习的"高性价比"更为凸显。学习强国 App 让学习的航道更加宽阔,让学生的未来更加精彩。请你模仿学习强国 App 的首页,制作一个仿学习强国 App 的首页。

第5章
常用API（1）

本章导读

本章主要讲解 uni-app 的常用 API，内容包括 API 概述、计时器、界面交互、网络、数据缓存、路由等，最后用一个智云翻译案例来演示各 API 的应用。

学习目标

知识目标	1. 掌握相应 API 的使用方法 2. 开发并完善智云翻译项目
能力目标	1. 具备熟练应用各种 API 的能力 2. 具备使用 Vuex 插件管理数据的能力 3. 具备项目测试能力 4. 具备使用插件的能力
素质目标	1. 具有团队协作精神 2. 具有良好的软件编码规范素养 3. 具备遵循软件项目开发流程的职业素养 4. 具有探索新知、不畏困难的精神

知识思维导图

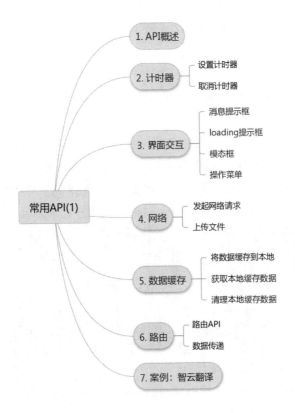

5.1 API 概述

在学习 uni-app 的 API 模块之前，先来了解什么是 API。API（Application Program Interface，应用程序接口）是指一些预先定义好的应用程序接口。API 在移动应用开发中扮演着重要的角色，基本上所有的移动应用程序都要用到 API 来实现相应功能。

uni-app 提供丰富的 API，包括基础、网络、路由、界面交互、数据缓存、媒体、设备、绘画、文件、第三方服务等模块。下面对常用模块分别进行介绍。

- 基础模块：包含日志输出、计时器、拦截器、base64 字符串和 ArrayBuffer 相互转换等工具 API。
- 网络模块：包含发送 HTTP 网络请求、文件上传下载、WebSocket、SocketTask、UDP 通信等与网络相关的 API。
- 路由模块：提供了实现页面路由、得到当前页面栈实例、页面通信、窗口动画等 API。
- 界面交互模块：用于界面交互的 API，如提示框、模态框，以及媒体查询、设置导航栏、设置 TabBar、下拉刷新、设置背景等各种与界面相关的 API。
- 数据缓存模块：提供了本地数据缓存的功能。

- 媒体模块：提供了图片、路由、文件、音频、视频、直播等各种媒体 API。
- 设备模块：提供了访问系统信息、内存、剪贴板，以及拨打电话、访问蓝牙设备等与设备相关的 API。
- 绘画模块：提供了有关操作 Canvas 的 API。
- 文件模块：包含保存文件、删除文件、得到文件信息、打开文件、得到文件系统管理器等 API。
- 第三方服务模块：包括获取服务供应商、登录、分享、支付、推送、发送语音等常见的 API。

上面这些 API 一般都通过 uni 全局对象调用，比如 uni.getSystemInfoSync。这些 API 中，一般用于监听事件的 API 都是以"on"开头的；用于得到数据的 API 都是以"get"开头的；用于设置数据的 API 都是以"set"开头的。大部分 API 都提供了同步和异步的两种操作。

许多 API 进行了 Promise 化，可以用 Promise 的调用方式来调用。示例代码如下。

```
const task = uni.connectSocket(    // 正常使用
 success(res){
  console.log(res)
 }
)
uni.connectSocket().then(res => {   // Promise 化
    console.log(res) // 此处的 res 为正常使用时 success 回调的 res
})
```

以下几类 API 没有进行 Promise 化：

（1）同步的方法（即以 Sync 结尾）：例如 uni.getSystemInfoSync；

（2）以 create 开头的方法：例如 uni.createMapContext；

（3）以 Manager 结尾的方法：例如 uni.getBackgroundAudioManager。

5.2 计时器

5.2.1 设置计时器

uni-app 提供了计时器 API，可以用两种方式设置计时器：使用 setTimeout 在计时到期以后执行注册的回调函数；使用 setInterval 按照指定的周期（以 ms 计）来执行注册的回调函数。

这两种方式的区别在于：setTimeout 是达到设定时间后执行一次计时器，如设置 5min 后执行计时器，那么在 5min 后计时器就启动并执行一次；setInterval 是按照设定的周期来执行计时器，如设置每 5min 执行一次计时器，那么每间隔 5min 计时器就会启动并执行一次。

1. setTimeout

setTimeout(callback, delay, rest)：设定一个计时器，在计时到期以后执行注册的回调函数。setTimeout 参数说明如表 5-1 所示。

表 5-1　setTimeout 参数说明

参数	类型	必填	说明
callback	Function	是	回调函数
delay	Number	否	执行回调函数的时间间隔，单位为 ms
rest	Any	否	param1、param2……paramN 等附加参数，它们会作为参数传递给回调函数

setTimeout 返回值为 Number 类型，表示计时器的编号。这个值可以传递给 clearTimeout 来取消计时。

2．setInterval

setInterval(callback, delay, rest)：设定一个计时器，按照指定的周期（以 ms 计）来执行注册的回调函数。delay 为两次执行回调函数的时间间隔，单位为 ms，其他参数与 setTimeout 的相同。

5.2.2　取消计时器

setTimeout 和 setInterval 设定的计时器都可以取消。clearTimeout(timeoutID)：取消由 setTimeout 设置的计时器。clearInterval(intervalID)：取消由 setInterval 设置的计时器。

参数说明：timeoutID 和 intervalID 为 Number 类型，是要取消的计时器的编号。

【实例 5-1】演示计时器的应用。

新建一个名为 uniappch05 的 uniapp 项目，选择 Vue2。新建 4 个页面，分别为 index.vue、three.vue、welcome.vue、timeout.vue。在 pages.json 中进行 tabBar 配置，设置 index.vue 和 three.vue 为 tabBar 页面（three.vue 在实例 5-8 中使用）。具体操作参考第 2 章的实例。在 pages.json 中设置 welcome.vue 页面为 pages 节点的第一项，让其作为应用启动页。

实例 5-1

注意：

在 uniappch05 项目中，在<template></template>模块中所用的 class 样式来源于 uni.css（将 1.4 节中 HelloTest_01 项目的 uni.css 文件复制到本项目）。

（1）主页 index.vue

在主页中设置导航，点击按钮跳转到对应的 API 页面，如图 5-1 所示。具体代码省略，与实例 2-9 类似。

（2）显示欢迎页

在 welcome.vue 页面中，显示欢迎页，2s 后跳转到 index.vue 页面。在 onLoad 函数中设置计时器，并读取系统信息，得到页面高度。在 pages.json 文件中，将 welcome.vue 作为 pages 节点的第一项。

图 5-1　主页运行效果

welcome.vue 的代码如下。

```
<template>
    <image src="../../static/image/welcome.webp" :style="{width:'100%', height: height +'px'}" mode="scaleToFill"></image>
</template>
<script>
    export default{
        data() {
            return {
                height:0
            }
        },
        onLoad() {
            var that = this
            uni.getSystemInfo({
                success:function (res) {
                    that.height = res.windowHeight

                }
            });
            setTimeout(function(){
                uni.switchTab({
                    url:'../index/index'
                })
            },2000)
        }
    }
</script>
<style> </style>
```

（3）演示计时器的启动与停止

点击按钮可启动或停止计时器。每个计时器开启后，只有取消后才能重新开启。timeout.vue 的具体代码如下。

```
<template>
    <view class="content">
        <view class="btnwrap">
            <button @click="setTimeout" :disabled="time1 != 0"> 开启计时器（5s）</button> <button @click="clearTimeout">取消计时器</button>
        </view>
        <view class="btnwrap">
            <button @click="setInterval" :disabled="time2 !=0"> 开启间隔计时器</button> <button @click="clearInterval">取消间隔计时器</button>
        </view>
    </view>
</template>
<script>
    export default {
        data() {
            return {
                time1:0,
                time2:0
```

```
            },
            methods:{
                setTimeout() {
                    this.time1 = setTimeout(function(res){
                        let data =  res;
                        console.log(data);
                    }, 5000, {
                        "message":"计时器——我和大自然有个约会"
                    }); //启动计时器
                },
                clearTimeout(){
                    clearTimeout(this.time1);
                    this.time1 = 0;
                },
                setInterval() {
                    this.time2 = setInterval(function(res){
                        console.log(res);}, 1000, {"message":"计时器——绿水青山"});
//启动计时器
                },
                clearInterval(){
                    clearInterval(this.time2);
                    console.log(this.time2);
                    this.time2 = 0;
                    console.log(this.time2);
                }
            }
        }
</script>
<style>
.btnwrap{
    display:flex;
    flex-direction:row;
    justify-content:space-around;
    padding:20px;
}
</style>
```

在微信开发者工具中的运行效果如图 5-2 所示。

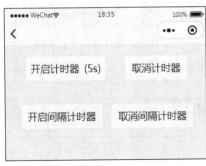
（a）应用页面

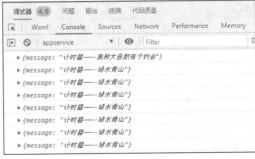
（b）控制台输出结果

图 5-2 计时器运行效果

5.3 界面交互

界面交互 API 包括 uni.showToast（显示消息提示框）API、uni.hideToast（隐藏消息提示框）API、uni.showModal（显示模态框）API、uni.showLoading（显示 loading 提示框）API、uni.hideLoading（隐藏 loading 提示框）API、uni.showActionSheet（显示操作菜单）API。使用这些界面交互 API，可以给用户提供良好的使用体验，提供提示信息。

5.3.1 消息提示框

uni.showToast(OBJECT)：显示消息提示框，其参数说明如表 5-2 所示。

uni.hideToast()：隐藏消息提示框。

表 5-2 uni.showToast(OBJECT)参数说明

参数	类型	必填	说明	平台差异
title	String	是	提示的内容，长度与 icon 的取值有关	
icon	String	否	图标，有效值如表 5-3 所示	
image	String	否	自定义图标的本地路径(App 平台暂不支持 GIF 图片)	App、H5、微信小程序、百度小程序支持
mask	Boolean	否	是否显示透明蒙层，防止触摸穿透，默认值为 false	App、微信小程序支持
duration	Number	否	提示的延迟时间，单位为 ms，默认值为 1500	
position	String	否	显示位置，填写有效值后只有 title 属性生效，且不支持通过 uni.hideToast 隐藏消息提示框。可取值为 top、center、bottom	App 支持

icon 的取值说明如表 5-3 所示。其中 error、fail、exception 的支持平台较少，这里不再列举。

表 5-3 icon 的取值说明

值	说明	平台差异
success	显示成功图标，此时 title 文本在小程序平台最多显示 7 个汉字的长度	支付宝小程序无显示长度限制
loading	显示加载图标，此时 title 文本在小程序平台最多显示 7 个汉字的长度	支付宝小程序不支持
none	不显示图标，此时 title 文本在小程序平台最多可显示两行，App 仅支持单行显示	

【实例 5-2】演示消息提示框的使用。

新建 Interactive.vue，然后在 index.vue 的数组 list 中添加 "{name: '交互反馈',url: 'Interactive' }"。在 Interactive.vue 页面中添加一组按钮，点击这些按钮弹出相应的消息提示框。

具体代码如下。

```html
<template>
    <view>
        <view class="uni-padding-wrap">
            <view class="uni-btn-v">
                <button type="default" @tap="toast1Tap">点击弹出默认消息提示框</button>
                <button type="default" @tap="toast2Tap">点击弹出设置duration的消息提示框</button>
                <button type="default" @tap="toast3Tap">点击弹出显示loading的消息提示框</button>
                <!-- #ifndef MP-ALIPAY -->
                <button type="default" @tap="toast4Tap">点击弹出显示自定义图片的消息提示框</button>
                <!-- #endif -->
                <button type="default" @tap="hideToast">点击隐藏消息提示框</button>
            </view>
        </view>
    </view>
</template>
<script>
    export default {
        data() {
            return { }
        },
        methods:{
            toast1Tap:function () {
                uni.showToast({
                    title:"默认，显示success"
                })
            },
            toast2Tap:function () {
                uni.showToast({
                    title:"延迟3s",
                    duration:3000
                })
            },
            toast3Tap:function () {
                uni.showToast({
                    title:"loading",
                    icon:"loading",
                    duration:5000
                })
            },
            toast4Tap:function () {
                uni.showToast({
                    title:"logo",
                    image:"../../static/logo.png"
                })
            },
            hideToast:function () {
                uni.hideToast()
            }
        }
    }
</script>
```

在微信开发者工具中的演示效果如图 5-3 所示。

（a）"默认，显示 success"的显示效果

（b）"延迟 3s"的显示效果

（c）"loading"的显示效果

（d）"logo"的显示效果

图 5-3 消息提示框的运行效果

5.3.2 loading 提示框

uni.showLoading (OBJECT)：用来显示 loading 提示框，其参数说明如表 5-4 所示。

uni.hideLoading()：用来隐藏 loading 提示框。

表 5-4 uni.showLoading (OBJECT)的参数说明

参数	类型	必填	说明	平台差异
title	String	是	提示的内容，显示在加载图标的下方	
mask	Boolean	否	是否显示透明蒙层，防止触摸穿透，默认值为 false	H5、App、微信小程序、百度小程序支持
success	Function	否	接口调用成功的回调函数	
fail	Function	否	接口调用失败的回调函数	
complete	Function	否	接口调用结束的回调函数（调用成功、失败都会执行）	

【实例 5-3】演示 loading 提示框的使用。

在 Interactive.vue 的<template></template>模块添加按钮。

```
<button class="btn-load" type="primary" @click="showLoading">显示 loading 提示框</button>
<button @click="hideLoading">隐藏 loading 提示框</button>
```

在<script></script>模块的 method 中添加方法。

```
hideLoading() {    uni.hideLoading();    },
showLoading() {
           uni.showLoading({    title:'loading'    });
}
```

弹出的 loading 提示框与实例 5-2 的 loading 提示框一样，如图 5-3（c）所示。

5.3.3 模态框

uni.showModal(OBJECT)：显示模态框，其中可以只有一个【确定】按钮，也可以同时有【确定】和【取消】按钮。类似于一个 API 整合了 HTML 中的 alert、confirm。其参数说明如表 5-5 所示。

表 5-5 uni.showModal(OBJECT)参数说明

参数	类型	必填	说明	平台差异
title	String	否	提示的标题	
content	String	否	提示的内容	
showCancel	Boolean	否	是否显示【取消】按钮，默认值为 true	
cancelText	String	否	【取消】按钮的文字，默认值为"取消"	
cancelColor	HexColor	否	【取消】按钮的文字颜色，默认值为"#000000"	H5、微信小程序、百度小程序支持
confirmText	String	否	【确定】按钮的文字，默认值为"确定"	
confirmColor	HexColor	否	【确定】按钮的文字颜色，在 H5 中的默认值为"#007aff"，在微信小程序中的默认值为"#576B95"，在百度小程序中的默认值为"#3c76ff"	H5、微信小程序、百度小程序支持
editable	Boolean	否	是否显示输入框	H5 和 App 3.2.10+、微信小程序 2.17.1+支持
placeholderText	String	否	显示输入框的提示文本	同 editable
success	Function	否	接口调用成功的回调函数，返回参数 res.confirm 为 true 时表示点击了【确定】按钮、res.cance 为 true 时，表示点击了【取消】按钮	

续表

参数	类型	必填	说明	平台差异
fail	Function	否	接口调用失败的回调函数	
complete	Function	否	接口调用结束的回调函数（调用成功、失败都会执行）	

【实例 5-4】演示模态框的使用。

在 Interactive.vue 的<template></template>模块中添加按钮。

```
<button type="primary" @tap="modalTap">点击弹出有标题模态框</button>
<button type="default" @tap="noTitlemodalTap">点击弹出无标题模态框</button>
```

在<script></script>模块的 method 中添加方法。

```
modalTap(e) {
    uni.showModal({
        title:"温馨提示",
        content:"为了您的账号安全，本次操作需要您进行登录。",
        showCancel:true,
        confirmText:"确定",
        success:function (res) {
            if (res.confirm) {
                console.log('用户点击确定');
            } else if (res.cancel) {
                console.log('用户点击取消');
            }
        }
    })
},
noTitlemodalTap(e) {
    uni.showModal({
     content:"本次交易存在风险，暂不能支付。请确认对方身份，如无误可 15 min 后再次发起交易。",
        confirmText:"确定",
        showCancel:false
    })
}
```

在微信开发者工具中运行，效果如图 5-4 所示。

（a）温馨提示的显示效果　　　　　　　　（b）用户点击"确定"按钮的显示效果

图 5-4　模态框运行效果

5.3.4　操作菜单

uni.showActionSheet(OBJECT)：从页面底部向上弹出操作菜单，其参数说明如表 5-6 所示。

表 5-6 uni.showActionSheet(OBJECT)参数说明

参数	类型	必填	说明	平台差异
title	String	否	菜单标题	App、H5、支付宝小程序、钉钉小程序、微信小程序 3.4.5+（仅真机有效）支持
alertText	String	否	警示文本	微信小程序（仅真机有效）支持
itemList	Array<String>	是	按钮的文字数组	微信、百度、字节跳动小程序中数组长度最大为 6
itemColor	HexColor	否	按钮的文字颜色，字符串格式，默认值为"#000000"	App-iOS、字节跳动小程序、飞书小程序不支持
popover	Object	否	大屏设备弹出原生选择按钮框的指示区域，默认居中显示。其有 top、left、width、height，分别用于指定坐标定位或宽度、高度	App-iPad 2.6.6+、H5 2.9.2 支持
success	Function	否	接口调用成功的回调函数，返回参数 tapIndex，表示用户点击的按钮的索引，从上到下，从 0 开始	
fail	Function	否	接口调用失败的回调函数	
complete	Function	否	接口调用结束的回调函数（调用成功、失败都会执行）	

【实例 5-5】演示操作菜单的使用。

在 Interactive.vue 的<template></template>模块添加按钮。

```
<button class="target" type="default" @tap="actionSheetTap">弹出操作菜单</button>
```

在<script></script>模块的 method 中添加方法。

```
actionSheetTap() {
    uni.showActionSheet({
        title:'选择学位',
        itemList:['学士','硕士','博士'],
        success:(e) => {
            console.log(e.tAPIndex);
            uni.showToast({
                title:"点击了第" + e.tAPIndex + "个选项",
                icon:"none"
            })
        }
    })
}
```

实例 5-2 至实例 5-5

在浏览器中运行的效果如图 5-5 所示。

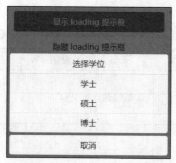

图 5-5 操作菜单运行效果

5.4 网络

uni-app 为开发者提供了一系列网络的 API，包括发起请求、上传和下载、WebSocket、UDP 通信。本节讲解发起请求中的发起网络请求 uni.request 和文件上传 uni.uploadFile。

5.4.1 发起网络请求

发起网络请求使用 uni.request(OBJECT)，相当于 AJAX 在实际开发中获取服务器端接口数据，其使用方式类似于 jQuery 的 AJAX。其参数说明如表 5-7 所示。

表 5-7 uni.request(OBJECT)参数说明

参数	类型	必填	默认值	说明	平台差异
url	String	是		开发者服务器接口地址	
data	Object/String/ArrayBuffer	否		请求的参数	App 3.3.7 以下不支持 ArrayBuffer 类型
header	Object	否		设置请求的头部，头部中不能设置 referer	App、H5 平台会自动带上 cookies，且 H5 平台不可手动修改
method	String	否	GET	有效值详见下文说明	
timeout	Number	否	60000	超时时间，单位为 ms	H5（HBuilderX 2.9.9+）、App（HBuilderX 2.9.9+）、微信小程序 2.10.0、支付宝小程序支持
dataType	String	否	json	如果设为 json，会尝试对返回的数据做一次 JSON.parse 处理	
responseType	String	否	text	设置响应的数据类型。有效值为 text、arraybuffer	支付宝小程序不支持

续表

参数	类型	必填	默认值	说明	平台差异
sslVerify	Boolean	否	TRUE	验证 SSL 证书	仅 App-Android（HBuilderX 2.3.3+）支持，不支持本地打包
withCredentials	Boolean	否	FALSE	跨域请求时是否携带凭证（cookies）	仅 H5（HBuilderX 2.6.15+）支持
firstIpv4	Boolean	否	FALSE	DNS 解析时优先使用 IPv4	仅 App-Android（HBuilderX 2.8.0+）支持
success	Function	否		收到开发者服务器成功返回的回调函数	
fail	Function	否		接口调用失败的回调函数	
complete	Function	否		接口调用结束的回调函数（调用成功、失败都会执行）	

参数说明如下。

（1）data：最终发送给服务器的 data 是 String 类型，如果传入的 data 不是 String 类型，会被转换成 String 类型。转换规则如下。

> 对于 GET 方法，会将数据转换为查询字符串。例如 { name: 'name', age: 18 } 转换后是 name=name&age=18。

> 对于 POST 方法且 header['content-type'] 为 application/json 的数据，会进行 JSON 序列化。

> 对于 POST 方法且 header['content-type'] 为 application/x-www-form-urlencoded 的数据，会将数据转换为查询字符串。

（2）method：可取值有 GET、POST、PUT、DELETETE、CONNECT、HEAD、OPTIONS、TRACE。App、H5、微信小程序支持所有可取值。具体有关平台的兼容性可以查看 uni-app 官网。其中较常用且兼容性较好的是 GET、POST。method 属性的取值取决于服务器端接口文档的描述，不得自己随意设置。

（3）success 的回调函数参数说明如表 5-8 所示。

表 5-8　success 的回调函数参数说明

参数	类型	说明
data	Object/String/ArrayBuffer	开发者服务器返回的数据
statusCode	Number	开发者服务器返回的 HTTP 状态码
header	Object	开发者服务器返回的 HTTP Response Header
cookies	Array.<string>	开发者服务器返回的 cookies，格式为字符串数组

注意：

在各个小程序平台运行时，使用网络相关的 API 前需要配置域名白名单。打开微信开发者工具,单击【详情】按钮,打开【本地设置】选项卡,勾选【不校验合法域名、web-view（业务域名）、TLS 版本以及 HTTPS 证书】复选框,如图 5-6 所示。

图 5-6　配置域名白名单

【实例 5-6】演示发起网络请求。

在 uniappch05 项目的 pages 文件夹下创建 request/request.vue 文件,在 request.vue 中添加代码。完整代码如下。

实例 5-6

```
<template>
    <view>  演示 request 网络请求！</view>
</template>
<script>
    export default {
        data() {
            return {
            }
        },
        onLoad() {
            uni.request({
                url:'https://unidemo.dcloud.net.cn/API/news',
                method:'GET',
                data:{},
                success:res => {
```

```
                    console.log("返回的数据",res);
                },
                fail:() => {},
                complete:() => {}
            })
        },
        methods:{
        }
    }
</script>
<style> </style>
```

在 Chrome 浏览器中浏览 request.vue 页面，打开开发者工具，查看控制台的输出。在控制台中可以看到请求成功的数据，如图 5-7 所示，其中包括 data 字段的值。

```
返回的数据                                                                        null:66
▼{data: Array(20), statusCode: 200, header: {…}, cookies: Array(0), errMsg: "request:ok"}
  ▶cookies: []
  ▶data: (20) [{…}, {…}, {…}, {…}, {…}, {…}, {…}, {…}, {…}, {…}, {…}, {…}, {…}, {…}, {…}, {…}, {…}…
   errMsg: "request:ok"
  ▶header: {content-type: "application/json", pragma: "no-cache"}
   statusCode: 200
  ▶__proto__: Object
at pages/request/request.vue:33
```

图 5-7　data 字段的值

uni-app 内部对 uni.request 进行了 Promise 封装，调用成功会进入 then 方法回调，调用失败会进入 catch 方法回调。

将实例 5-6 中的 onLoad 用下列代码替换。

```
onLoad() {
    uni.request({
        url:'https://unidemo.dcloud.net.cn/API/news',
        method:'GET',
        data:{}
    }).then((res) => {
        console.log("返回的数据", res);
    })
}
```

在使用 POST 请求时，必须配置 header 部分，往往会配置其 content-type 属性，该属性值取决于后端接口的 content-type 类型。注册用户，示例代码如下。

```
let URL = "http://localhost:3004" + "/users";
uni.request({
    url:URL,
    data:{
        phoneNum:this.phoneNum,
        nickName:this.nickName,
        password:this.password
        },
        method:'POST',
        header:{
            'content-type':'application/x-www-form-urlencoded',
        },
        success:function(res) {
            console.log("成功", res)
```

```
        },
        fail:function(res) {
            console.log("失败",res)
        }
    });
```

5.4.2 上传文件

使用 uni.uploadFile(OBJECT)可以将本地资源上传到开发者服务器。由客户端发起一个 POST 请求，其中 content-type 为 multipart/form-data。在真机中测试上传文件时，需要配置域名白名单；在微信开发者工具中演示时，需勾选【不校验合法域名、web-view（业务域名）、TLS 版本以及 HTTPS 证书】复选框即可使用 uni.uploadFile(OBJECT)，如图 5-6 所示。其调用格式为：

```
uni.uploadFile(OBJECT);
```

其参数说明如表 5-9 所示。

表 5-9 uni.uploadFile(OBJECT)参数说明

参数	类型	必填	说明	平台差异
url	String	是	开发者服务器 URL	
files	Array	是	需要上传的文件列表。使用 files 时，filePath 和 name 不生效	App、H5 2.6.15+支持
fileType	String		文件类型，有效值为 image、video、audio	仅支付宝小程序支持，且必填
file	File	否	要上传的文件对象	仅 H5 2.6.15+支持
filePath	String	是	要上传文件资源的路径	
name	String	是	文件对应的 key，开发者在服务器端通过这个 key 可以获取文件的二进制内容	
header	Object	否	HTTP 请求 header，头部中不能设置 referer	
timeout	Number	否	超时时间，单位为 ms	H5、App（HBuilderX 2.9.9+）支持
formData	Object	否	HTTP 请求中其他的 表单数据	
success	Function	否	接口调用成功的回调函数	
fail	Function	否	接口调用失败的回调函数	
complete	Function	否	接口调用结束的回调函数（调用成功、失败都会执行）	

success 回调函数中有两个参数：data、statusCode。其含义与 uni.request 的 success 回调函数一致。

页面通过 uni.chooseImage 等接口获取一个本地资源的临时文件路径后，可通过此接口将本地资源上传到指定服务器。

【实例 5-7】演示图片或视频文件上传。

在本例中,实现点击【选择图片】view 组件,触发 click 事件,执行 chooseImage 方法。在 chooseImage 方法中,先调用 uni.chooseImage 打开文件,如果调用成功,则调用 uni.uploadFile 上传文件,如果上传成功,则弹出消息提示框。如果在微信小程序平台运行,则可打开相册选择图片。完整的代码如下。

实例 5-7

```
<template>
    <view>
        <view class="common-page-head">
            <view class="common-page-head-title">{{title}}</view>
        </view>
        <view class="uni-padding-wrap uni-common-mt uni-common-mb">
            <view class="demo">
                <block v-if="imageSrc">
                    <image :src="imageSrc" class="image" mode="widthFix"></image>
                </block>
                <block v-else>
                    <view class="uni-hello-addfile" @click="chooseImage">+ 选择图片</view>
                </block>
            </view>
        </view>
    </view>
</template>
<script>
    export default {
        data() {
            return {
                title:'上传图片',
                imageSrc:''
            }
        },
        onUnload() {
            this.imageSrc = '';
        },
        methods:{
            chooseImage:function() {
                uni.chooseImage({
                    count:1,
                    sizeType:['compressed'],
                    sourceType:['album'],
                    success:(res) => {
                        console.log('chooseImage success, temp path is', res.tempFilePaths[0])
                        var imageSrc = res.tempFilePaths[0]
                        uni.uploadFile({
                            url:'https://unidemo.dcloud.net.cn/upload',
                            filePath:imageSrc,
                            fileType:'image',
                            name:'data',
                            success:(res) => {
                                console.log('uploadImage success, res is:', res)
```

```
                            uni.showToast({
                                title:'上传成功',
                                icon:'success',
                                duration:1000
                            })
                            this.imageSrc = imageSrc
                        },
                        fail:(err) => {
                            console.log('uploadImage fail', err);
                            uni.showModal({
                                content:err.errMsg,
                                showCancel:false
                            });
                        }
                    });
                },
                fail:(err) => {
                    console.log('chooseImage fail', err)
                    // #ifdef MP
                    uni.getSetting({
                        success:(res) => {
                            let authStatus = res.authSetting['scope.album'];
                            if (!authStatus) {
                                uni.showModal({
                                    title:'授权失败',
                                    content:'Hello uni-app需要从您的相册获取图片,请在设置界面打开相关权限',
                                    success:(res) => {
                                        if (res.confirm) {
                                            uni.openSetting()
                                        }
                                    }
                                })
                            }
                        }
                    })
                    // #endif
                }
            })
        }
    }
}
</script>
<style>
    .image {
        width:60%;
    }
    .demo {
        background:#FFF;
        padding:50rpx;
    }
    .uni-hello-addfile {
        text-align:center;
```

```
        line-height:200rpx;
        background:#FFF;
        padding:50rpx;
        margin-top:10px;
        font-size:38rpx;
        color:#808080;
    }
</style>
```

在 H5 平台运行的效果如图 5-8 所示。

（a）选择图片的显示效果　　　　　　　（b）上传成功的显示效果

图 5-8　图片上传效果

5.5 数据缓存

uni-app 提供了数据本地缓存功能，如可以将用户信息缓存到本地，这样就不用每次调用服务器来获取这些信息。数据缓存 API 就是用来将需要的数据保存到本地的，另外还可以用于获取本地缓存数据、移除缓存数据及清理缓存数据。在实际开发中，数据缓存 API 经常用于保存会员登录状态信息、购物车、历史记录等数据。

5.5.1　将数据缓存到本地

uni-app 为数据缓存到本地提供了两种方式：uni.setStorage(OBJECT)和 uni.setStorageSync (KEY,DATA)。

1. uni.setStorage(OBJECT)

uni.setStorage(OBJECT)以异步方式将数据存储在本地缓存指定的 key 中，会覆盖掉 key 原来对应的内容。其参数说明如表 5-10 所示。

表 5-10　uni.setStorage(OBJECT)参数说明

参数	类型	必填	说明
key	String	是	本地缓存中指定的 key
data	Any	是	需要缓存的内容，只支持原生类型及能够通过 JSON.stringify 序列化的对象

续表

参数	类型	必填	说明
success	Function	否	接口调用成功的回调函数
fail	Function	否	接口调用失败的回调函数
complete	Function	否	接口调用结束的回调函数（调用成功、失败都会执行）

【实例 5-8】 以异步方式存储对象的示例代码如下。

```
<script>
    export default {
        onLoad() {
        var user = this.getUserInfo();
        console.log(user);
        uni.setStorage({
            key:'user',
            data:user,
            success:function(res) {
                console.log(res);
            }
        })
        },
        methods:{
            getUserInfo() {
                var user = new Object();
                user.id = '10000'
                user.nickname = '小甜鱼';
                user.username = 'smallfish'
                return user;
            }
        }
    }
</script>
```

在微信开发者工具的调试器的【Storage】选项卡里可以查看缓存的数据，如图 5-9 所示。

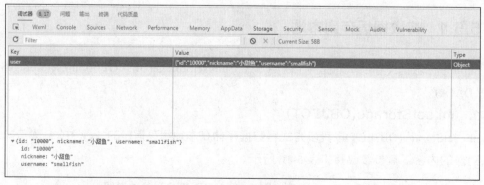

图 5-9　查看缓存的数据

2. uni.setStorageSync(KEY,DATA)

uni.setStorageSync(KEY,DATA)以同步方式将数据存储在本地缓存指定的 key 中，会覆盖掉 key 原来对应的内容，比异步缓存数据更简单一些。其参数说明如表 5-11 所示。

表 5-11　uni.setStorageSync(KEY,DATA)参数说明

参数	类型	必填	说明
key	String	是	本地缓存中指定的 key
data	Any	是	需要缓存的内容，只支持原生类型及能够通过 JSON.stringify 序列化的对象

示例代码如下。

```
onLoad() {
    var user = this.getUserInfo();
    console.log(user);
    uni.setStorage('user', use)   ;
}
```

同样可以在微信开发者工具的调试器的【Storage】选项卡中查看缓存数据。

3．异步和同步的差别

将数据缓存到本地，不管是以同步方式还是以异步方式，都是通过"key/value"的形式存储数据的，只不过同步方式需要等本地缓存成功后，才可以继续执行下面的程序，而异步方式则不需要等待本地缓存成功就可以继续执行下面的程序。在数据缓存比较耗时的情况下，可以使用异步方式进行数据缓存，保证程序不用等待就能继续执行。

5.5.2　获取本地缓存数据

uni-app 为获取本地缓存数据提供了 4 个 API：
- uni.getStorage(OBJECT)：以异步方式从本地缓存中获取指定 key 对应的内容；
- uni.getStorageSync(KEY)：以同步方式从本地缓存中获取指定 key 对应的内容；
- uni.getStorageInfo()：以异步方式获取本地所有 key 的集合；
- uni.getStorageInfoSync()：以同步方式获取本地所有 key 的集合。

前两个 API 用于从指定 key 里获得缓存数据，而后两个 API 用于获取本地所有 key 的集合。

1．uni.getStorage(OBJECT)

uni.getStorage(OBJECT)使用异步方式从本地缓存中获取指定 key 对应的内容。

其参数说明如表 5-12 所示。

表 5-12　uni.getStorage（OBJECT）参数说明

参数	类型	必填	说明
key	String	是	本地缓存中指定的 key
success	Function	否	接口调用成功的回调函数，res 为{data:key 对应的内容}
fail	Function	否	接口调用失败的回调函数
complete	Function	否	接口调用结束的回调函数（调用成功、失败都会执行）

在 5.5.1 小节中，我们使用 uni.setStorage(OBJECT)将 user 以异步方式保存到本地，下面使用 uni.getStorage 来获取本地缓存数据，具体代码如下。

```
onLoad(){    //以异步方式获取本地缓存数据
    uni.getStorage({
      key:'user',
      success:function(res){
        console.log(res);
      }
    })
}
```

2. uni.getStorageSync(KEY)

uni.getStorageSync(KEY)是一个同步的 API，只有一个参数 key，用来从本地缓存中以同步方式获取指定 key 对应的内容。

参数说明：key 为本地缓存中指定的 key。

示例代码如下：

```
onLoad(){    //以同步方式获取本地缓存数据
  var userSync = uni.getStorageSync('userSync');
  console.log(userSync);
}
```

3. uni.getStorageInfo(OBJECT)

uni.getStorageInfo(OBJECT)以异步方式获取本地所有 key 的集合，其参数说明如表 5-13 所示。

表 5-13　uni.getStorageInfo（OBJECT）参数说明

参数	类型	必填	说明
success	Function	是	接口调用成功的回调函数，详见表 5-14
fail	Function	否	接口调用失败的回调函数
complete	Function	否	接口调用结束的回调函数（调用成功、失败都会执行）

success 回调函数参数说明如表 5-14 所示。

表 5-14　success 回调函数参数说明

参数	类型	说明
keys	Array<String>	当前缓存中所有的 key
currentSize	Number	当前占用的空间大小，单位为 KB
limitSize	Number	限制的空间大小，单位为 KB

示例代码如下。

```
onLoad(){
  uni.getStorageInfo({
    success:function(res){
      console.log(res);
    }
  })
}
```

获取到本地所有的 key 后，根据 key 调用 uni.getStorage 或 uni.getStorageSync 接口就可

以获取本地缓存数据了。

4. uni.getStorageInfoSync()

uni.getStorageInfoSync()以同步方式获取本地所有 key 的集合，示例代码如下。

```
onLoad(){
  var storage = uni.getStorageInfoSync();
  console.log(Storage);
}
```

它和 uni.getStorageInfo(OBJECT)一样，都是在获取本地所有的 key 后根据 key 再查找完整的数据。

5.5.3 清理本地缓存数据

uni.removeStorage(OBJECT)、uni.removeStorageSync(KEY)用来从本地缓存中移除指定key；uni.clearStorage()、uni.clearStorageSync()用来清理本地缓存数据。

1. uni.removeStorage(OBJECT)

uni.removeStorage(OBJECT)用来以异步方式从本地缓存中移除指定的 key。其参数说明类似于 uni.getStorage(OBJECT)，有 key、success、fail、complete。

下面从本地缓存中移除 key 值为 user 的数据，具体代码如下。

```
onLoad(){    //异步移除 key 值为 user 的数据
    uni.removeStorage({
      key:'user',
        success:function(res){
          console.log(res);
      }
  })
}
```

移除完后，在本地缓存列表里就找不到 key 值为 user 的缓存数据了，表示移除成功。

2. uni.removeStorageSync(KEY)

uni.removeStorageSync(KEY)用来以同步方式从本地缓存中移除指定的 key，它的效果和 uni.removeStorage(OBJECT)一样。

示例代码如下。

```
onLoad(){    //同步移除 key 值为 userSync 的数据
    uni.removeStorageSync('userSync');
}
```

3. uni.clearStorage()和 uni.clearStorageSync()

uni.clearStorage()和 uni.clearStorageSync()用来清理本地所有缓存数据，前者采用异步方式，后者采用同步方式。

示例代码如下。

```
uni.clearStorage();

try {
    uni.clearStorageSync()
```

```
            } catch(e) {
            }
```

【实例 5-9】演示数据存储、读取和清理。

在本例中，演示使用同步或异步方式存储或读取不同类型的数据，以及清理本地缓存数据。在微信开发者工具中的演示效果如图 5-10 所示。输入 key 和 value 的值后，点击【存储数据】按钮保存数据，或者点击【清理数据】按钮清理数据，可以在【Storage】选项卡中看到变化。在 onLoad 中存储 Object 类型的数据，点击【异步读取 onLoad 函数中存储的对象】或【同步读取 onLoad 函数中存储的对象】，分别以异步或同步方式读取对象数据。

实例 5-8 和
实例 5-9

（a）输入数据

（b）存储数据

（c）【Storage】选项卡的内容

图 5-10 数据存储的演示效果

具体代码如下。

```
<template>
    <view>
        <view class="common-page-head">
            <view class="common-page-head-title">{{title}}</view>
        </view>
        <view class="uni-list">
            <view class="uni-list-cell">
                <view class="uni-list-cell-left">
                    <view class="uni-label">key</view>
                </view>
                <view class="uni-list-cell-db">
```

```html
                <input class="uni-input" type="text" placeholder="请输入 key" name="key" :value="key"
                    @input="keyChange" />
                </view>
            </view>
            <view class="uni-list-cell">
                <view class="uni-list-cell-left">
                    <view class="uni-label">value</view>
                </view>
                <view class="uni-list-cell-db">
                    <input class="uni-input" type="text" placeholder="请输入 value" name="data" :value="data"
                        @input="dataChange" />
                </view>
            </view>
        </view>
        <view class="uni-padding-wrap">
            <view class="uni-btn-v">
                <button type="primary" class="btn-setstorage" @tap="setStorage">存储数据</button>
                <button @tap="getStorage">读取数据</button>
                <button @tap="clearStorage">清理数据</button>
                <button @tap="getStorageObject">异步读取 onLoad 函数中存储的对象</button>
                <button @tap="getStorageObjectSys">同步读取 onLoad 函数中存储的对象</button>
            </view>
        </view>
        <view v-if="flag" class=" uni-padding-wrap uni-bg-green" >
            <view class="uni-list-cell uni-comment-content">用户编号:{{user.id}}</view>
            <view class="uni-list-cell uni-comment-content">用户昵称:{{user.nickname}}</view>
            <view class="uni-list-cell uni-comment-content">用户名:{{user.username}}</view>
        </view>
    </view>
</template>
<script>
    export default {
        data() {
            return {
                title:"数据存储",
                key:'',
                data:'',
                flag:false,
                user:{  id:'',
                    nickname:'',
                    username:''
                }
            }
        },
```

```
onLoad() {
    var user = this.getUserInfo();
    console.log(user);
    uni.setStorage({
        key:'user',
        data:user,
        success:function(res) {
            console.log(res);
        }
    })

},
methods:{
    keyChange:function(e) {
        this.key = e.detail.value
    },
    dataChange:function(e) {
        this.data = e.detail.value
    },
    getUserInfo() {
        var user = new Object();
        user.id = '10000'
        user.nickname = '小甜鱼';
        user.username = 'smallfish'
        return user;
    },
    getStorageObject() {
        uni.getStorage({
            key:'user',
            success:function(res) {

                console.log( res.data);
                uni.showModal({
                    title:'读取数据成功',
                    content:"data:'" + JSON.stringify(res.data) + "'",
                    showCancel:false
                })
            }
        })
    },
    getStorageObjectSys() {
        this.user = uni.getStorageSync("user");
        this.flag = true;
        console.log(this.user);
    },
    getStorage:function() {
        var key = this.key,
            data = this.data;
        if (key.length === 0) {
            uni.showModal({
                title:'读取数据失败',
                content:"key 不能为空",
                showCancel:false
```

```
                })
            } else {
                uni.getStorage({
                    key:key,
                    success:(res) => {
                        uni.showModal({
                            title:'读取数据成功',
                            content:"data:'" + res.data + "'",
                            showCancel:false
                        })
                    },
                    fail:() => {
                        uni.showModal({
                            title:'读取数据失败',
                            content:"找不到 key 对应的数据",
                            showCancel:false
                        })
                    }
                })
            }
        },
        setStorage:function() {
            var key = this.key
            var data = this.data
            if (key.length === 0) {
                uni.showModal({
                    title:'保存数据失败',
                    content:'key 不能为空',
                    showCancel:false
                })
            } else {
                uni.setStorage({
                    key:key,
                    data:data,
                    success:(res) => {
                        uni.showModal({
                            title:'存储数据成功',
                            // #ifndef MP-ALIPAY
                            content:JSON.stringify(res.errMsg),
                            // #endif
                            // #ifdef MP-ALIPAY
                            content:data,
                            // #endif
                            showCancel:false
                        })
                    },
                    fail:() => {
                        uni.showModal({
                            title:'存储数据失败',
                            showCancel:false
                        })
                    }
```

```
                })
            }
        },
        clearStorage:function() {
            try{
            uni.clearStorageSync()
            this.key = '',
                this.data = ''
            uni.showModal({
                title:'清理数据成功',
                content:' ',
                showCancel:false
            })
            }catch(e){
                Console log('清理数据失败');
            }
        }
    }
}
</script>
<style>
    .btn-setstorage {
        background-color:#007aff;
        color:#ffffff;
    }
</style>
```

5.6 路由

uni-app 可以在页面中设置导航，使用 navigator 组件实现页面跳转，也可以在 JS 文件里通过路由 API 进行页面跳转。

5.6.1 路由 API

表 5-15 中列举了 navigator 组件的 open-type 属性的 5 种取值，分别对应 5 种不同的跳转方式。这些跳转方式都有与之对应的 API。

表 5-15 open-type 属性的 5 种取值及 API

属性值	API	说明
navigate	uni.navigateTo(OBJECT)	保留当前页面，跳转到应用内的某个页面，使用 uni.navigateBack 返回
redirect	uni.redirectTo(OBJECT)	关闭当前页面，跳转到应用内的某个页面
switchTab	uni.switchTab(OBJECT)	跳转到 tabBar 页面，并关闭其他所有非 tabBar 页面
reLaunch	uni.reLaunch(OBJECT)	关闭所有页面，打开应用内的某个页面
navigateBack	uni.navigateBack(OBJECT)	关闭当前页面，返回上一级页面或多级页面

uni.navigateTo(OBJECT)、uni.navigateBack (OBJECT)API 的 OBJECT 参数说明如表 5-16 所示。

表 5-16　OBJECT 参数说明（1）

参数	类型	必填	默认值	说明	差异说明
url	String	是		需要跳转的应用内非 tabBar 页面的路径	uni.navigateTo 中的参数
delta	Number	否	1	返回的页面数，如果 delta 值大于现有页面数，则返回首页	uni.navigateBack 的参数
animationType	String	否	pop-in	窗口显示的动画效果，详见 uni-app 官网的"窗口动画"	App
animationDuration	Number	否	300	窗口动画持续时间，单位为 ms	App
events	Object	否		页面间通信接口，用于监听被打开页面发送到当前页面的数据	uni.navigateTo 的参数
success	Function	否		接口调用成功的回调函数	
fail	Function	否		接口调用失败的回调函数	
complete	Function	否		接口调用结束的回调函数（调用成功、失败都会执行）	

uni.redirectTo(OBJECT)、uni.switchTab(OBJECT)、uni.reLaunch(OBJECT)API 的 OBJECT 参数说明如表 5-17 所示。

表 5-17　OBJECT 参数说明（2）

参数	类型	必填	说明	差异说明
url	String	是	需要跳转的应用内非 tabBar 页面的路径	uni.switchTab 的路径不能带参数
success	Function	否	接口调用成功的回调函数	
fail	Function	否	接口调用失败的回调函数	
complete	Function	否	接口调用结束的回调函数（调用成功、失败都会执行）	

其中，url 为需要跳转的应用内页面的路径，路径后可以带参数。参数与路径之间使用"?"分隔，参数名与参数值用"="相连，不同参数用"&"分隔；如"path?key=value&key2=value2"。如果要跳转的页面是 tabBar 页面，则路径后不能带参数。

uni.navigateTo(OBJECT)的 success 回调函数参数 res 的属性如表 5-18 所示。

表 5-18　res 的属性

属性	类型	说明	差异说明
eventChannel	EventChannel	和打开的页面进行通信	uni.navigateTo 的 success 回调函数

注意：
- uni.navigateTo、uni.redirectTo 只能打开非 tabBar 页面；
- 页面跳转有层级限制，不能无限制跳转新页面，微信小程序有 10 层限制；
- uni.switchTab 只能打开 tabBar 页面；
- uni.reLaunch 可以打开任意页面；
- 页面底部的 tabBar 由页面决定，即只要定义的是 tabBar 页面，底部都有 tabBar；
- 不能在 App.vue 里进行页面跳转；
- H5 平台页面刷新之后页面栈会消失，此时 uni.navigateBack 不能返回，如果一定要返回，可以使用 history.back 导航到浏览器的其他历史记录。

【实例 5-10】演示页面路由。

在 uniappch05 项目的 pages 文件夹下新建一个 navigate.vue 页面。在该页面中放置 5 个按钮，点击这些按钮分别演示 5 个路由 API。这里跳转的页面都是前面实例的页面。navigate.vue 的具体代码如下。

```
<template>
    <view>
        <view class="uni-padding-wrap">
            <view class="uni-btn-v">
                <button @tap="navigateTo">跳转至图片上传页</button>
                <button @tap="navigateBack">返回上一页</button>
                <button @tap="redirectTo">关闭当前页，跳到计时器页</button>
                <button @tap="switchTab">切换到 tabBar 页面</button>
                <button @tap="reLaunch">关闭所有页面，进入欢迎页</button>
            </view>
        </view>
    </view>
</template>
<script>
    export default {
        data() {
            return {}
        },
        methods:{
            navigateTo() {
                uni.navigateTo({
                    url:'../uploadfile/uploadfile'
                })
            },
            navigateBack() {
                uni.navigateBack();
            },
            redirectTo() {
                uni.redirectTo({
                    url:'../timeout/timeout'
                });
            },
            switchTab() {
                uni.switchTab({
```

```
            url:'../three/three'
        });
    },
    reLaunch() {
        uni.reLaunch({
            url:'../welcome/welcome'
        });
    }
  }
}
</script>
<style> </style>
```

5.6.2 数据传递

uni.navigateTo、uni.redirectTo、uni.reLaunch 这 3 个路由 API 的参数 url 都可以携带参数，如 "path?key=value&key2=value2"。uni-app 中接收路由参数与 Vue 不同，uni-app 是在 onLoad 生命周期函数中接收参数。onLoad 接收的参数为 Object 类型，会序列化上级页面传递的参数。

【实例 5-11】演示数据传递。

在本例中跳转新页面时会携带参数。在 navigate 文件夹下新建页面 acceptData.vue。

实例 5-10 和
实例 5-11

在 navigate.vue 的 <template></template> 模块添加按钮。

```
<button @tap="navigateToPage">跳转页面，并传递参数</button>
```

在 <style></style> 模块的 methods 中添加方法。

```
navigateToPage() {
            uni.navigateTo({
                url:"./acceptData/acceptData?id=100&nickname='rose'&username='张三'"
            })
        }
```

acceptData.vue 的代码如下。

```
<template>
  <view class="uni-padding-wrap">
      <text>{{obj}}</text>
  </view>
</template>
<script>
  export default {
      data() {
          return { obj:null  }
      },
      onLoad(object){
          this.obj = object;
          console.log(this.obj)
      },
      methods:{
      }
  }
</script>
<style></style>
```

运行项目，通过点击按钮进入 acceptData.vue 页面，acceptData.vue 会显示传递过来的数据，如图 5-11 所示。

注意：

url 参数有长度限制，太长的字符串会传递失败，这个问题可以采用全局变量或页面通信的方式来解决，具体参考 uni-app 官网的页面通信相关内容。如参数中出现空格等特殊字符时，需要对参数进行编码，这时可以使用 encodeURIComponent()。

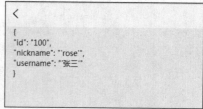

图 5-11 数据传递

```
uni.navigateTo{ //item 的内容为含有特殊字符的字符串
Url:"'/pages/test/test?item='+ encodeURIComponent(JSON.stringify(item'))
}

// 在 test.vue 页面接收参数
onLoad:function (option) {
  const item = JSON.parse(decodeURIComponent(option.item));
}
```

跳转到 tabBar 页面时是不能传递参数的，但是实际开发中需要在跳转到 tabBar 页面时传递参数。可以利用 uni.setStorageSync() 将参数存入本地缓存来实现传参。

5.7 案例：智云翻译

本案例利用有道智云 AI 开放平台实现中英文的翻译。在图 5-12 所示的页面中输入要翻译的内容，点击【翻译】按钮，进入翻译结果页，如图 5-13 所示。本案例还将实现查看历史翻译记录的功能，如图 5-14 所示。本例中使用扩展组件 uni-ui。

案例：智云翻译

图 5-12 翻译页

（a）单词翻译结果页

（b）句子翻译结果页

图 5-13　翻译结果页

图 5-14　历史翻译记录页

⚙ 实现步骤

1. 初始配置

（1）在有道智云 AI 开放平台注册账号并登录，点击头像进入控制台，点击应用总览页面中的【创建应用】按钮，进入图 5-15 所示的界面。

图 5-15　创建应用

创建应用成功后，可以得到该应用的 ID 和密钥。

（2）新建一个使用 uni-ui 的 uni-app 项目 translate-demo（注意选择 Vue2）。删除 pages

下的 index 文件夹，新建 3 个页面：翻译页 translate.vue、翻译结果页 result.vue、历史翻译记录页 history.vue。参照 2.1 节的内容实现 tabBar 导航。其中"翻译"tabBar 页面对应 translate.vue 页面，"历史"tabBar 页面对应 history.vue 页面，进入程序首页为 translate.vue 页面。pages.json.vue 文件内容如下。

```
{
  "pages":[    {
        "path" :"pages/translate/translate",
        "style" : {
            "navigationBarTitleText":"智云翻译",
            "navigationBarTextStyle":"white",
            "navigationBarBackgroundColor":"#4a90e2",
            "enablePullDownRefresh":false
        }
    } ,{
        "path" :"pages/history/history",
        "style" :     {
            "navigationBarTitleText":"智云翻译",
            "navigationBarTextStyle":"white",
            "navigationBarBackgroundColor":"#4a90e2",
            "enablePullDownRefresh":false
        }
    } ,{
        "path" :"pages/result/result",
        "style" :     {
            "navigationBarTitleText":"",
            "enablePullDownRefresh":false
        }
    } ],
  "globalStyle":{
      "navigationBarTextStyle":"black",
      "navigationBarTitleText":"uni-app",
      "navigationBarBackgroundColor":"#F8F8F8",
      "backgroundColor":"#F8F8F8",
      "app-plus":{"background":"#efeff4"}
  },
  "tabBar":{
      "backgroundColor":"#F8F8F8",
      "borderStyle":"black",
      "list":[{
         "iconPath":"static/tabbar/translate-icon.png",
         "selectedIconPath":"static/tabbar/translate-selected-icon.png",
         "pagePath":"pages/translate/translate",
         "text":"翻译"
      }, {
         "iconPath":"static/tabbar/history-icon.png",
         "selectedIconPath":"static/tabbar/history-selected-icon.png",
         "pagePath":"pages/history/history",
         "text":"历史"
      }]
  }
}
```

2. 跨域配置

打开 manifest.json 文件，在源代码视图的末尾添加 H5 的跨域配置，注意每项与前面的项需用"，"间隔。

```
"h5":{
    "devServer":{
        "proxy":{
            "/API":{
                "target":"http://openAPI.youdao.com",
                "changeOrigin":true, //是否跨域
                "secure":true, // 设置支持 HTTPS 的代理
                "pathRewrite":{
                    "^/youdaoAPI":"/"
                }
            }
        }
    }
}
```

在 main.js 文件中，根据 Vue 的版本，将下列代码添加到对应的位置。

```
// #ifdef H5
Vue.prototype.baseUrl="/youdaoAPI"
// #endif
// #ifndef H5
Vue.prototype.baseUrl="http://openAPI.youdao.com"
// #endif
```

3. 配置 Vuex

uni-app 中自带 Vuex 插件，不需要独立安装，可以直接使用。在项目中新建一个 store 文件夹，在此文件夹下新建 index.js 文件，输入以下代码。

```
import Vue from 'vue'
import Vuex from 'vuex'
Vue.use(Vuex)
const store = new Vuex.Store({    })    //实例化 store 对象
export default store    //导出 store 对象
```

在 main.js 文件中添加图 5-16 所示的代码，将 store 对象注册到全局。

4. CSS 样式的配置

本案例在新建时使用了 uni-ui，在 App.vue 中已经进行了配置。这里除了需要使用 uni-ui 的 index.scss 之外，还需要使用第 1 章的 HelloTest_01 项目中的 common/uni.css 样式文件。因此需要将该文件复制到 common 文件夹。在 App.vue 中的 <style></style> 中添加代码 "@import './common/uni.css';"，修改 page 的背景色为 aliceblue。具体代码如下。

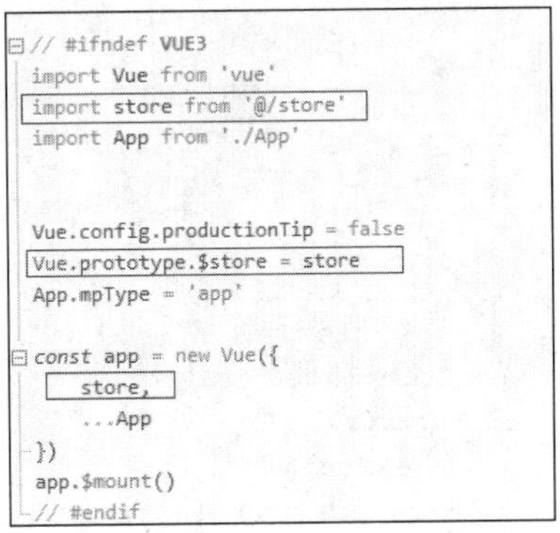

```
// #ifdef VUE3
import Vue from 'vue'
import store from '@/store'
import App from './App'

Vue.config.productionTip = false
Vue.prototype.$store = store
App.mpType = 'app'

const app = new Vue({
    store,
    ...App
})
app.$mount()
// #endif
```

图 5-16　在 main.js 中注册 store 对象

```
<style lang="scss">
    /*每个页面公共 CSS */
    @import './common/uni.CSS';
    @import '@/uni_modules/uni-scss/index.scss';
    /* #ifndef APP-NVUE */
    @import '@/static/customicons.css';
    // 设置整个项目的背景色
    page {      background-color:aliceblue;   }
    /* #endif */
    .example-info {
        font-size:14px;
        color:#333;
        padding:10px;
    }
</style>
```

5. translate.vue 页面的搭建

具体代码如下。

```
<template>
    <view class="container ">
        <view class="translate-content">
            <uni-row class="uni-common-mb">
                <uni-col :span="10" class="lefttext"><text>源语言</text></uni-col>
                <uni-col :span="12">
                    <uni-data-select v-model="valueSou" :localdata="languageList" :clear="false" @change="">   </uni-data-select>
                </uni-col>
            </uni-row>
            <uni-row class="uni-common-mb">
                <uni-col :span="10" class="lefttext"><text>目标语言</text></uni-col>
                <uni-col :span="12">
    <uni-data-select v-model="valueDec" :localdata="languageList" :clear="false" @change=""> </uni-data-select>
                </uni-col>
            </uni-row>
            <uni-row class="uni-flex  uni-ma-5">
    <textarea v-model="searchValue" placeholder="请输入要翻译的内容" class="txtcolor uni-flex-item "></textarea>
            </uni-row>
            <uni-row class="uni-flex uni-row uni-flex-item  ">
                <button type="primary" @click="clear">重置</button>
                <button type="primary" @click="translate">翻译</button>
            </uni-row>
        </view>
    </view>
</template>
<script>
    export default {
        data() {
            return {
                searchValue:'',
```

```
            valueSou:'zh-CHS',
            valueDec:'en',
            languageList:[{  value:'zh-CHS',  text:'简体中文'  },
                    {  value:'zh-CHT',  text:'繁体中文'  },
                    {  value:'en',  text:'英文'  },
                    {  value:'ja',  text:'日文'  },
                    {  value:'ko',  text:'韩文'  },
                    {  value:'fr',  text:'法文'  },
                    {  value:'es',  text:'西班牙文'}  ]
        }
    },
    methods:{
        clear(){    this.searchValue='';     }
    }
}
</script>
<style>
    .container {
        padding:20px;
        font-size:14px;
        line-height:24px;
        height:100%;
    }
    .lefttext {   text-align:center;   }
    .txtcolor {
        background-color:white;
        padding:20rpx;
        border:1rpx solid #e1e1e1;
    }
    .flex-grow {  flex-grow:1;}
</style>
```

6．实现文本翻译功能

在 translate.vue 中，编写【翻译】按钮单击方法 translate，在 translate 方法中，需要访问有道智云 API。实现步骤如下。

（1）安装 js-sha256。在 HBuilderX 中打开终端，运行命令 npm install js-sha256，结果如图 5-17 所示。

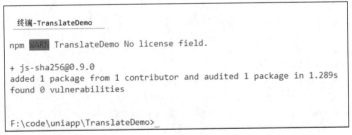

图 5-17　安装 js-sha256

（2）读者可以自行在有道智云 AI 开放平台中查看实现文本翻译功能的 JS 示例代码，如图 5-18 所示。

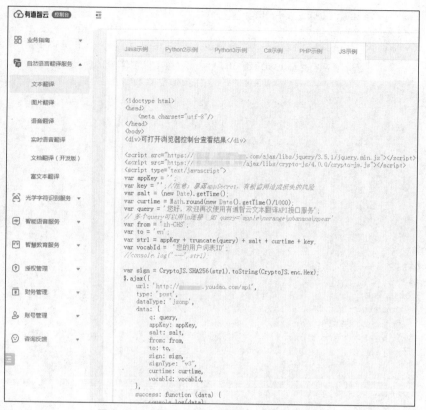

图 5-18　实现文本翻译功能的 JS 示例代码

（3）根据示例代码，编写 translate 方法。在 index.vue 的<script></script>部分添加相关代码。

```
<script>
    import { sha256 } from 'js-sha256'
    export default {
        data() {    ……省略部分与前面相同
        },
        methods:{
            clear() {   this.searchValue = ''      },
            truncate(q) {
                var len = q.length;
                if (len <= 20) return q;
                return q.substring(0, 10) + len + q.substring(len - 10, len);
            },
            translate() {
                let appKey = '224590bdb002e1d6';
                let key = 'Bei1koQl0kNRJp9RovOdG***********';  //读者需使用自己的 key
                let salt = (new Date).getTime();
                let curtime = Math.round(new Date().getTime() / 1000);
                let query = this.searchValue;
                let from = this.valueSou;
                let to = this.valueDec;
                let str1 = appKey + this.truncate(query) + salt + curtime + key;
                let sign = sha256(str1);
```

```
                uni.request({
                    //非 H5 时，url:'http://openAPI.youdao.com/API',
                    //H5 时,url:"/youdaoAPI/API",
                    url:this.baseUrl + "/API",
                    type:'post',
                    data:{
                        q:query,
                        appKey:appKey,
                        salt:salt,
                        from:from,
                        to:to,
                        sign:sign,
                        signType:"v3",
                        curtime:curtime
                    },
                    header:{
                        'custom-header':'hello'  //自定义请求头信息
                    },
                    success:(res) => {
                        console.log(res);
                    }
                })
            }
        }
    }
</script>
```

（4）运行 index.vue 页面，在 textarea 组件中输入"中国"，可以在控制台看到 res.data.dict 中 translation 属性的值为翻译的结果"China"，在 web 属性中有一组关联词组翻译，如图 5-19 所示。由于官网 API 时常更新，请求得到的结果可能有变化，均以官网 API 技术文档说明为准。

图 5-19　控制台输出结果

7. 在翻译结果页 result.vue 中显示翻译结果

result.vue 需要 index.vue 中的翻译结果。这里将翻译结果存入 store 对象里面，同时需要存储此次翻译的记录。

（1）编写 store/index.js 代码。

```js
import Vue from 'vue'
import Vuex from 'vuex'
Vue.use(Vuex)
//实例化 store 对象
const store = new Vuex.Store({
    state:{
        outcome:{},
        historyList:[],
    },
    getters:{
        getOutcome(state){
            return state.outcome;
        },
        getHistoryList(state){
            return state.historyList;
        }
    },
    mutations:{
        setOutcome(state,payload){
            state.outcome = payload;
        },
        addHistoryList(state,payload){
            state.historyList.push(payload);
        },
        saveData(state,payload){
            uni.setStorage({
                key:'outcome',
                data:state.outcome
            });
            uni.setStorage({
                key:'historyList',
                data:state.historyList
            })
        },
        readData(state,payload){
            let storageOutcome = uni.getStorageSync('outcome');
            state.outcome = storageOutcome ? storageOutcome :{};
            let storageHistory = uni.getStorageSync('historyList');
            state.historyList = storageHistory ? storageHistory :[];
        }
    }
})
export default store    //导出 store 对象
```

（2）修改 index.vue 中的 translate，请求成功后，将翻译结果存入 store 对象和缓存，然后导航至翻译结果页 result.vue。

```js
success:(res) => {
    // 将翻译结果对象存入 store 对象，便于跨页面使用
    this.$store.commit('setOutcome', res);
    this.$store.commit('addHistoryList', res);
```

```
            console.log(res);
            this.$store.commit('saveData');//将数据存入缓存
            uni.navigateTo({
               url:'/pages//result/result'
            });     }
```

8. 显示翻译结果

在 index.vue 页面中将 API 的返回结果存储在 store 对象的 outcome 中。进入页面并读取存储的值。在翻译结果页中,点击播放图标按钮 ◀»,将会播放音频。speak 方法用来实现此功能。<script></script>部分的代码如下。

```
    <script>
        export default {
            data() {
                return {
                    dataBean:this.$store.state.outcome.data ? this.$store.state.
outcome.data :this.$store.state.historyList[
                        this.$store.state.historyList.length - 1].data
                }
            },
            onShow() {
                this.dataBean = this.$store.state.outcome.data ? this.$store.state.
outcome.data :
                    this.$store.state.historyList[this.$store.state.historyList.
length - 1].data;
            },
            methods:{
                speak(url) {
                    let audioContext = uni.createInnerAudioContext();
                    audioContext.src = url;
                    audioContext.play();
                }
            }
        }
    </script>
```

index.vue 页面使用 uni-ui 的折叠面板来显示"词性变化""网络释义"。翻译分为单词翻译和句子翻译。两种翻译结果分别使用 v-if、v-else 来显示。该页面中使用的样式类部分来源于 uni.css 文件。

```
    <template>
        <view>
            <view class="uni-common-mt uni-common-pl" v-if="dataBean.isWord">
                <uni-row class="uni-center ">
                    <text>{{dataBean.query}}</text>
                </uni-row>
                <uni-row>
                    <uni-col :span="8"><text>英:{{dataBean.basic['us-phonetic']}}
</text></uni-col>
                    <uni-col :span="4">
        <uni-icons type="sound-filled" size="16" @click="speak(dataBean.basic
['us-speech'])"></uni-icons>
                    </uni-col>
                    <uni-col :span=" 8">
```

```
                <text>美:{{dataBean.basic['uk-phonetic']}}</text>
            </uni-col>
            <uni-col :span="4">
<uni-icons type="sound-filled" size="16" @click="speak(dataBean.basic['uk-speech'])"></uni-icons>
            </uni-col>
        </uni-row>
        <view v-for="(explain, index) in dataBean.basic.explains" :key="index" class="word-explain-field">
            <text>{{explain}}</text>
        </view>
        <uni-section title="其他" type="line" style="margin-right:30rpx;">
            <uni-collapse>
                <uni-collapse-item title="词性变化" :show-animation="true"
                    v-if="dataBean.basic.wfs && dataBean.basic.wfs.length>0">
                    <view class="content">
                        <view v-for="(item, index) in dataBean.basic.wfs" :key="index">
                            <uni-row>
                                <uni-col :span="2"></uni-col>
                                <uni-col :span="10">{{item.wf.name}}</uni-col>
                                <uni-col :span="12">{{item.wf.value}}</uni-col>
                            </uni-row>
                        </view>
                    </view>
                </uni-collapse-item>
                <uni-collapse-item title="网络语义" :show-animation="true" v-if="dataBean.web && dataBean.web.length>0">
                    <view class="content">
                        <view v-for="(item, index) in dataBean.web" :key="index">
                            <uni-row>
                                <uni-col :span="2"><text></text></uni-col>
                                <uni-col :span="10">{{item.key}}</uni-col>

                                <uni-col :span="12">
            <text v-for="(valueitem,i) in item.value" :key="i">{{valueitem}} </text>
                                </uni-col>
                            </uni-row>
                        </view>
                    </view>
                </uni-collapse-item>
            </uni-collapse>
        </uni-section>
    </view>
    <view v-else class="uni-common-mt uni-common-pl ">
        <uni-row class="uni-common-mb">
            <uni-col :span="6">
                <text>源语言</text>
            </uni-col>
            <uni-col :span="4">
<uni-icons type="sound-filled" size="16" @click="speak(dataBean.speakUrl)"></uni-icons>
```

```html
                </uni-col>
                <uni-col :span="4"></uni-col>
                <uni-col :span="6">          <text>目标语言</text>      </uni-col>
                <uni-col :span="4">
        <uni-icons type="sound-filled" size="16" @click="speak(dataBean.tSpeakUrl)"></uni-icons>
                </uni-col>
            </uni-row>
            <view class=" trans-result " style="">{{dataBean.translation[0]}}</view>
            <view class="" v-if="dataBean.web && dataBean.web.length>0">
                <uni-section title="其他" type="line" style="margin-right:30rpx;">
                    <uni-collapse ref="collapse">
                        <uni-collapse-item title="网络语义" :show-animation="true">
                            <view class="content">
                                <view v-for="(item, index) in dataBean.web" :key="index">
                                    <text> {{item.key}}</text>
                                    <view v-for="(valueItem, vIndex) in item.value" :key="vIndex"
                                        style="white-space:pre;color:#3c9cff;">
                                        <text>{{valueItem}}{{vIndex!=item.value.length-1 ? ', ' :''}}</text>
                                    </view>
                                </view>
                            </view>
                        </uni-collapse-item>
                    </uni-collapse>
                </uni-section>
            </view>
        </view>
    </view>
</template>
<script>
    export default {
        data() {
            return {
                dataBean:this.$store.state.outcome.data ? this.$store.state.outcome.data :this.$store.state.historyList[this.$store.state.historyList.length - 1].data
            }
        },
        onShow() {
            this.dataBean = this.$store.state.outcome.data ? this.$store.state.outcome.data :
                this.$store.state.historyList[this.$store.state.historyList.length - 1].data;
        },
        methods:{
            speak(url) {
                let audioContext = uni.createInnerAudioContext();
                audioContext.src = url;
                audioContext.play();
            }
```

```
        }
    }
</script>
<style>
    .content {
        padding:20rpx;
        height:auto;
    }
    .trans-result {
        margin-right:30rpx;
        padding:20rpx;
        border-radius:8rpx;
        background-color:white;
        margin-bottom:20rpx;
    }
    .word-explain-field {
        font-size:32rpx;
        padding-block:10rpx;
        color:#666666;
    }
</style>
```

9. 显示历史翻译记录

用 uni-ui 中的列表显示历史翻译记录,当点击列表项时,打开 result.vue 页面,显示翻译结果。

```
<template>
    <view>
        <uni-list >
            <uni-list-item  v-for="(item,index) in historyList" :key="index" showArrow
         clickable :title="item.data.query" :rightText="getLanguageTransfer(item.data.l)" @click="toDetail(item)" />
        </uni-list>
    </view>
</template>
<script>
    export default {
        data() {
            return {
                historyList:[],
                languageList:[{   value:'zh-CHS',       text:'简中'    },
                    {   value:'zh-CHT',    text:'繁中'},
                    {   value:'en',    text:'英'    },
                    {   value:'ja',    text:'日'    },
                    {   value:'ko',    text:'韩'    },
                    {   value:'fr',    text:'法'    },
                    {   value:'es',    text:'西班牙'    }]
            }
        },
        methods:{
            toDetail(historyItem) {
```

```
                this.$store.commit('setOutcome', historyItem);
                uni.navigateTo({
                    url:'/pages/result/result'
                });
            },
            getLanguageTransfer(languageStr) {
                let languages = languageStr.split('2');
                let fromStr = this.languageList.find(item => {
                    return item.value == languages[0];
                });
                let toStr = this.languageList.find(item => {
                    return item.value == languages[1];
                });
                return fromStr.text + '->' + toStr.text;
            }
        },
        onShow() {
            this.historyList = Array.from(this.$store.state.historyList);
            this.historyList.reverse();
        }
    }
</script>
<style>    </style>
```

10. 完善程序

当要翻译之前已经翻译过的内容，则不需要重新访问有道智云 AI 开放平台，可以直接从历史翻译记录中获取翻译结果。在翻译页 index.vue 中的 translate 方法中，添加判定代码。如果 searchValue、valueSou、valueDec，与历史翻译记录中的某条记录一致，则直接从历史翻译记录中获取翻译结果。

```
translate() {
    //判断之前是否进行过此翻译
    let searchHistory = this.$store.state.historyList ? this.$store.state.historyList :[];;
    let itemLocal = searchHistory.find(item => {
        let queryEqual = item.data.query == this.searchValue;
        let itemLanguages = item.data.l.split('2');
        return queryEqual && this.fromLanguage == itemLanguages[0] && this.toLanguage == itemLanguages[
            1];
    });
    if (itemLocal) { // 若进行过则从本地获取翻译结果
        store.commit('setOutcome', itemLocal);
        uni.navigateTo({
            url:'/pages/result/result'
        })
    } else {
        let appKey = '224590bdb002e1d6';
        ……    //此处省略 translate 方法之前的代码
    }
}
```

本章小结

本章主要通过实例介绍了计时器、界面交互、发起网络请求、文件上传、数据缓存、路由等 API。API 是指一些预先定义好的应用程序接口，开发者可以通过 API 快速实现相关功能。熟练掌握 API 的用法，可以让开发者开发出功能强大的移动应用。最后通过一个智云翻译的综合案例，提升读者的 API 应用能力。

项目实战

使用百度 AI 开放平台完成一个植物识别的页面，页面效果如图 5-20 所示。

图 5-20　植物识别页面效果

拓展实训项目

党的二十大报告指出"推进文化自信自强，铸就社会主义文化新辉煌"。中国电影作为中国文化事业的重要组成部分，承担着相应的文化责任和时代使命。请参考豆瓣以及各电影平台开发"电影之家"小程序，其主要功能有热门电影推荐、电影搜索、电影详情、电影评论等。小程序数据来源可以为开源的豆瓣影评接口。

第6章
常用API（2）

本章导读

本章主要讲解 uni-app 的常用 API，包括媒体控制、文件操作、设备操作、登录等 API；最后通过一个仿网易云音乐 App 的音乐播放器案例，综合应用本章的 API。

学习目标

知识目标	1. 掌握常用 API 的使用方法 2. 开发并完善音乐播放器项目 3. 掌握自定义组件的方法
能力目标	1. 具备熟练应用常用 API 的能力 2. 具备开发并使用自定义组件的能力 3. 具备项目测试能力
素质目标	1. 具有团队协作精神 2. 具有良好的软件编码规范素养 3. 具有独立思考、分析问题、解决问题的能力 4. 提升音乐素养，树立正确的审美观、人生观和价值观

知识思维导图

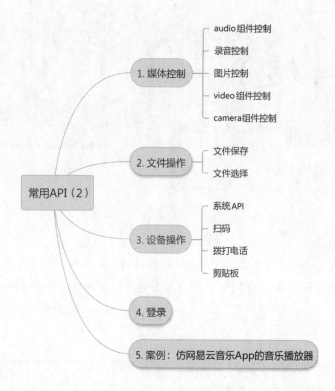

6.1 媒体控制

在移动应用中经常需要实现音频播放、视频播放、相机拍照、实时音视频播放、实时音视频录制等功能,可以通过 audio(音频)组件及音频 API 实现音频播放功能;通过 video(视频)组件及视频 API 实现视频播放功能;通过 camera(相机)组件及相机 API 实现相机拍照功能;通过 liver-player 组件实现实时音视频播放功能;通过 liver-pusher 组件实现实时音视频录制功能。本节将详细介绍常用的媒体组件及媒体 API 的应用。

6.1.1 audio 组件控制

uni-app 主要通过 uni.createInnerAudioContext()和 uni.getBackgroundAudioManager()实现 audio 组件控制。

1. uni.createInnerAudioContext()

对应音频的使用,一般建议使用内部 audio 上下文 innerAudioContext 对象。通过 uni.createInnerAudioContext()能够创建并返回 innerAudioContext 对象。该对象的属性如表 6-1 所示,其方法如表 6-2 所示。

表 6-1　innerAudioContext 对象的属性

属性	类型	说明	只读	平台差异
src	String	音频的数据链接，用于直接播放	否	微信小程序不支持本地路径
startTime	Number	音频开始播放的位置（单位为 s），默认值为 0	否	
autoPlay	Boolean	音频是否自动开始播放，默认值为 false	否	
loop	Boolean	音频是否循环播放，默认值为 false	否	
obeyMuteSwitch	Boolean	音频是否遵循系统静音开关的控制，当此属性值为 false 时，即使用户打开了静音开关，音频也能继续发出声音，默认值为 true	否	微信、百度、字节跳动、飞书、京东、快手（仅 iOS）小程序支持
duration	Number	当前音频的长度（单位为 s），src 合法时返回，通过 onCanplay 获取	是	
currentTime	Number	当前音频的播放位置（单位为 s，小数），src 合法时返回，保留小数点后 6 位	是	
paused	Boolean	当前音频是否处于暂停或停止状态，true 表示暂停或停止，false 表示正在播放	是	
buffered	Number	音频缓冲的时间点，仅保证当前播放时间点到指定时间点内容已缓冲	是	
volume	Number	音频音量。范围为 0~1	否	
sessionCategory	String	音频播放模式	否	App 3.3.7+支持
playbackRate	Number	音频播放的倍速。可取值为 0.5、0.8、1.0、1.25、1.5、2.0，默认值为 1.0	否	App 3.4.5+、微信小程序 2.11.0、支付宝、字节跳动 2.33.0+、快手、百度 3.120.2+等小程序支持

补充说明如下。

sessionCategory 的可取值：

Ambient——不中止其他音频播放，不能后台播放音频，静音后无声音；

soloAmbient——中止其他音频播放，不能后台播放音频，静音后无声音；

Playback——中止其他音频播放，可以后台播放音频，静音后有声音，其默认值为"playback"。

表 6-2　innerAudioContext 对象的方法

方法	参数	说明
play		播放（H5 平台部分浏览器需在用户交互时播放）
pause		暂停播放
stop		停止播放

续表

方法	参数	说明
seek	position	跳转到指定位置，单位为 s
destroy		销毁当前实例
onCanplay、offCanplay	callback	音频进入可播放状态事件，但不保证后面可以流畅播放；offCanplay 用于取消监听 onCanplay 事件
onPlay、offPlay	callback	onPlay：音频播放事件；offPlay 用于取消监听 onPlay 事件
onPause、offPause	callback	onPause：音频暂停事件；offPause 用于取消监听 onPause 事件
onStop、offStop	callback	onStop：音频停止事件；offStop 用于取消监听 onStop 事件
onEnded、offEnded	callback	onEnded：音频自然播放结束事件；offEnded 用于取消监听 onEnded 事件
onTimeUpdate、offTimeUpdate	callback	onTimeUpdate：音频播放进度更新事件；offTimeUpdate 用于取消监听 onTimeUpdate 事件
onError、offError	callback	onError：音频播放错误事件；offError 用于取消监听 onError 事件
onWaiting、offWaiting	callback	onWaiting：音频加载中事件，当音频因为数据不足，需要停下来加载时会触发该事件；offWaiting 用于取消监听 onWaiting 事件
onSeeking、offSeeking	callback	onSeeking：音频进行查询操作事件；offSeeking 用于取消监听 onSeeking 事件
onSeeked、offSeeked	callback	onSeeked：音频完成查询操作事件；offSeeked 用于取消监听 onSeeked 事件

补充说明如下。

（1）以 off 开头的取消监听的方法，支持的平台有微信小程序 1.9.0+、支付宝小程序、字节跳动小程序、百度小程序。

（2）errCode：表示错误代码，系统错误代码为 10001、网络错误代码为 10002、文件错误代码为 10003、格式错误代码为 10004、未知错误代码为-1。

（3）iOS 支持 M4A、WAV、MP3、AAC、AIFF、CAF、WAV 格式的音频；Android 支持 FLAC、M4A、OGG、APE、AMR、WMA、WAV、MP3、MP4、AAC 格式的音频。为了保证兼容性和音质，推荐使用 MP3 格式的音频。

【实例 6-1】演示音频 API。

演示播放一个音频，效果如图 6-1 所示。

图 6-1 音频播放效果

实现步骤

（1）新建一个 uni-app 项目 uniappch06，该项目与 uniappch05 项目类似，其中创建了一个导航页，用于进入各个页面。

实例 6-1

（2）复制 1.4 节的示例项目 HelloTest_01 中 common 文件夹下的 util.js 到本项目的 common 文件夹下。这里需要用到 util.js 文件中的 formatTime 方法。该方法用于将时间整数值转换为"hh:mm:ss"样式。该方法定义如下。

```
function formatTime(time) {
    if (typeof time !== 'number' || time < 0) {
        return time
    }
    var hour = parseInt(time / 3600)
    time = time % 3600
    var minute = parseInt(time / 60)
    time = time % 60
    var second = time
    return ([hour, minute, second]).map(function(n) {
        n = n.toString()
        return n[1] ? n :'0' + n
    }).join(':')
}
```

（3）新建 audio.vue 页面，在 pages.json 文件中将 audio.vue 页面的标题设置为 false，代码如下。

```
"pages":[    {
        "path":"pages/audio/audio",
        "style":{    "navigationBarTitleText":"uni-app",
            "app-plus":{"titleNView":false}
        }
    }    ],
```

（4）修改 pages/audio/audio.vue，代码如下。

```
<template>
    <view class="content" :style="{height:windowHeight}">
        <view class="title">  <text>中国心-------来源:网易云音乐</text>  </view>
        <view class="player-songpic">
            <image src="../../static/中国心.png" mode="widthFix"></image>
        </view>
        <view class="player-slider">
            <view class="player-currentTime">  {{currentTimeStr}}  </view>
<slider class="slider" activeColor="#b6b6b6" backgroundColor="#dedede" block-size="12" :value="position"  :min="0" :max="duration" @changing="onchanging" @change="onchange"></slider>
            <view class="player-duration">    {{durationStr}}    </view>
        </view>
            <view class="play-bar">    <!-- playbar -->
                <image class="opt" src="../../static/back.png" @tap="sub()"></image>
                <image class="opt2" :src="playImage" @click="play"></image>
                <image class="opt" src="../../static/next.png" @tap="add()"></image>
            </view>
    </view>
</template>
<script>
    import * as util from '../../common/util.js' //导入 util.js 中的方法，并取别//名为 util
```

```js
        const audioUrl = 'http://music.163.com/song/media/outer/url?id=
1392457251.mp3'; //播放歌曲的地址
        export default {
            data() {
                return {
                    windowHeight:'100px',//窗体的高度在 onLoad 中获取
                    isPlaying:false,//是否正在播放
                    isPlayEnd:false,//是否处于结尾
                    currentTime:0, //当前时间
                    duration:100,//总的时间
                    durationStr:'00:00:00', //总时间字符串
                    currentTimeStr:'00:00:00',           //当前时间字符串
                }
            },
            computed:{
                position() {
                    this.currentTimeStr = util.formatTime(Math.floor(this.currentTime));
                    return this.isPlayEnd ? 0 :this.currentTime;
                },
                playImage() {
                    return this.isPlaying ? "/static/pause.png" :"/static/play.png"
                }
            },
            onLoad() {
                this._isChanging = false;  //播放进度是否改变
                this._audioContext = null;
                uni.getSystemInfo({  //调用获取系统信息的 API，得到窗体的高度
                    success:(res) => {
                        this.windowHeight = res.windowHeight + "px";
                    }
                });
                this.createAudio(); //调用方法创建 innerAudioContext 对象
            },
            onUnload() {
                if (this._audioContext != null && this.isPlaying) {
                    this.stop();
                }
            },
            methods:{
                createAudio() {
                    var innerAudioContext = this._audioContext = uni.createInnerAudioContext();
                    innerAudioContext.autoplay = false;
                    innerAudioContext.src = audioUrl;
                    innerAudioContext.onPlay(() => {
                        console.log('开始播放');
                    });
                    innerAudioContext.onTimeUpdate((e) => {
                        if (this._isChanging === true) {
                            return;
                        }
```

```
            this.currentTime = innerAudioContext.currentTime || 0;
            this.duration = innerAudioContext.duration || 0;
        });
        innerAudioContext.onEnded(() => {
            this.currentTime = 0;
            this.isPlaying = false;
            this.isPlayEnd = true;
        });
        innerAudioContext.onError((res) => {
            this.isPlaying = false;
            console.log(res.errMsg);
            console.log(res.errCode);
        });
        return innerAudioContext;
    },
    onchanging() { //拖动过程中执行
        this._isChanging = true;

    },
    onchange(e) { //完成一次拖动后触发执行
        console.log(e.detail.value);
        console.log(typeof e.detail.value);
        this._audioContext.seek(e.detail.value);
        this._isChanging = false;
    },
    play() { //点击中间的播放暂停图标按钮后执行
        var duration = this._audioContext.duration;
        this.duration = duration;
        this.durationStr =util.formatTime(Math.floor(this.duration));
        if (this.isPlaying) { //如果正在播放音频，则暂停
            this.pause();
            return;
        }
        this.isPlaying = true; //如果当前处于停止状态，播放音频
        this._audioContext.play();
        this.isPlayEnd = false;
    },
    pause() {
        this._audioContext.pause();
        this.isPlaying = false;
    },
    stop() {//关闭页面时调用
        this._audioContext.stop();
        this.isPlaying = false;
    },
    add(){ //前进
        this._audioContext.seek(this.currentTime +10);
    },
    sub(){ //后退
        this._audioContext.seek(this.currentTime -10);
    }
}
```

```
        }
    </script>
    <style>
        .content {
            background-image:linear-gradient(to    bottom,   #2893d0,   #5c5f36,
#2893d0);
            display:flex;
            flex-direction:column;
            align-items:center;
        }
        .title {
            color:#FFFFFF;
            margin:100rpx;
            font-size:40rpx;
        }
        .player-songpic{
            width:90%;
            display:flex;
            flex-direction:row;
            justify-content:center;
            position:absolute;
            top:30%;
        }
        .player-songpic image {
            width:60%;
        }
        .player-slider {
            width:95%;
            height:50rpx;
            display:flex;
            align-items:center;
            justify-content:space-between;
            position:absolute;
            bottom:250rpx;
        }
        .player-slider .player-currentTime,
        .player-slider .player-duration {

            width:100rpx;
            height:100%;
            font-size:10px;
            color:#c1c1c1;
            line-height:44rpx;
        }
        .slider {
            flex:1;
        }
        .play-opt-bar {
            width:80%;
            height:60rpx;
            display:flex;
            align-items:center;
            justify-content:space-between;
            padding:0 70rpx;
```

```
        font-size:34px;
        position:absolute;
        bottom:320rpx;
    }
    .play-bar {
        width:70%;
        display:flex;
        align-items:center;
        justify-content:space-between;
        padding:0 70rpx;
        font-size:30px;
        position:absolute;
        bottom:120rpx;
    }
    .opt {
        width:64rpx;
        height:64rpx;
    }
    .opt2 {
        width:100rpx;
        height:100rpx;
    }
    .player {
        width:650rpx;
        height:50rpx;
        display:flex;
        align-items:center;
        justify-content:space-between;
        position:absolute;
        bottom:220rpx;
    }
</style>
```

2. uni.getBackgroundAudioManager()

通过 uni.getBackgroundAudioManager()可获取全局唯一的背景音频管理器 backgroundAudioManager 对象。背景音频不是背景音乐,而是类似 QQ 音乐 App,在后台运行时仍然可以播放的音频。如果不需要在 App 后台运行时继续播放音频,那么不应该使用本 API,而应该使用普通音频的 uni.createInnerAudioContext()。

uni.getBackgroundAudioManager()支持的平台有 App、微信小程序、百度小程序、字节跳动小程序、QQ 小程序、京东小程序。

backgroundAudioManager 对象的属性如表 6-3 所示。

表 6-3　backgroundAudioManager 对象的属性

属性	类型	说明	只读
duration	Number	当前音频的长度(单位为 s),只有在当前有合法的 src 时返回	是
currentTime	Number	当前音频的播放位置(单位为 s),只有在当前有合法的 src 时返回	是
paused	Boolean	当前音频是否处于暂停或停止状态,true 表示暂停或停止,false 表示正在播放	是

续表

属性	类型	说明	只读
src	String	音频的数据源,默认为空字符串,当设置了新的 src 时,会自动开始播放音频,目前支持的格式有 M4A、AAC、MP3、WAV	否
startTime	Number	音频开始播放的位置(单位为 s)	否
buffered	Number	音频缓冲的时间点,仅保证当前播放时间点到指定时间点的内容已缓冲	是
title	String	音频标题,用作原生音频播放器的音频标题。通过原生音频播放器中的分享功能,分享出去的卡片标题,也将使用该值	否
epname	String	专辑名,通过原生音频播放器中的分享功能,分享出去的卡片简介,也将使用该值	否
singer	String	歌手名,通过原生音频播放器中的分享功能,分享出去的卡片简介,也将使用该值	否
coverImgUrl	String	封面图 URL,用作原生音频播放器背景图。通过原生音频播放器中的分享功能,分享出去的卡片配图及背景,也将使用该图	否
webUrl	String	页面链接,通过原生音频播放器中的分享功能,分享出去的页面链接,也将使用该值	否
protocol	String	音频协议。默认值为"http",设置为 "hls" 时可以支持播放 HLS 协议的直播音频,App 平台暂不支持	否
playbackRate	Number	播放的倍速。可取值为 0.5、0.8、1.0、1.25、1.5、2.0,默认值为 1.0。(App 3.4.5+、微信基础库 2.11.0+、支付宝小程序、字节跳动小程序 2.33.0+、快手小程序、百度小程序 3.120.2+ 支持)	否

backgroundAudioManager 对象的方法与 InnerAudioContext 对象的方法几乎一致,只是没有以 off 开头的取消监听的方法。

示例代码如下。

```
const bgAudioManager = uni.getBackgroundAudioManager();
bgAudioManager.title = '致爱丽丝';
bgAudioManager.singer = '暂无';
bgAudioManager.coverImgUrl = 'https://bjetxgzv.cdn.bspapp.com/VKCEYUGU-uni-app-doc/7fbf26a0-4f4a-11eb-b680-7980c8a877b8.png';
bgAudioManager.src = 'https://bjetxgzv.cdn.bspapp.com/VKCEYUGU-hello-uniapp/2cc220e0-c27a-11ea-9dfb-6da8e309e0d8.mp3';
```

6.1.2 录音控制

录音通常用于聊天系统中的发送语音或者录音直接转文字功能。uni.getRecorderManager() 可以获得全局唯一的录音管理器 recorderManager 对象,支持的平台有 App、微信小程序、百度小程序、字节跳动小程序、飞书小程序、QQ 小程序、京东小程序。

recorderManager 对象的方法如表 6-4 所示。

表 6-4　recorderManager 对象的方法

方法	参数	说明	平台差异
start	options	开始录音	
pause		暂停录音	App 暂不支持
resume		继续录音	App 暂不支持
stop		停止录音	
onStart	callback	录音开始事件	
onPause	callback	录音暂停事件	
onStop	callback	录音停止事件，会回调文件地址	
onResume	callback	监听继续录音事件	
onInterruptionBegin	callback	监听录音因为受到系统占用而被中断开始事件	微信小程序、百度小程序、QQ 小程序支持
onInterruptionEnd	callback	监听录音中断结束事件。收到此事件之后小程序内暂停的录音才可再次录音成功	微信小程序、百度小程序、QQ 小程序支持
onFrameRecorded	callback	已录制完指定帧大小的文件，会回调录音分片结果数据。如果设置了 frameSize，则会回调此事件	App 暂不支持
onError	callback	录音错误事件，会回调错误信息	

补充说明如下。

onInterruptionBegin 事件在以下场景会触发：微信语音聊天、微信视频聊天、QQ 语音聊天、QQ 视频聊天、电话响铃、接听电话。此事件触发后，录音会被暂停。pause 事件在此事件后触发。

recorderManager 对象的 start(options)的参数说明如表 6-5 所示。

表 6-5　recorderManager 对象的 start(options)的参数说明

参数	类型	必填	说明	支持平台
duration	Number	否	指定录音的时长，单位为 ms；如果传入了合法的 duration，到达指定的 duration 后自动停止录音，最大值为 600000ms（10 min），默认值为 60000ms（1 min）	
sampleRate	Number	否	采样率，有效值为 8000、16000、44100	
numberOfChannels	Number	否	录音通道数，有效值为 1、2	小程序支持
encodeBitRate	Number	否	编码码率，码率具体有效值参考 uni-app 官网	小程序支持
format	String	否	音频格式，有效值为 aac、mp3、wav、PCM。App 默认值为 mp3，小程序默认值为 aac	App、小程序支持
frameSize	String	否	指定帧大小，单位为 KB。传入 frameSize 后，在录制指定帧大小的内容后，会回调录制的文件内容。若不指定则不回调。仅支持 MP3	App、百度小程序不支持

【实例 6-2】演示录音控制。

在微信小程序中演示录音、播放录音、停止录音等功能。在操作之前，请在 manifest.json 的【微信小程序配置】中输入微信开发者工具登录的微信小程序 AppID。实例演示效果如图 6-2 所示。

实例 6-2

（a）初始状态

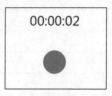
（b）录音中

（c）停止后

（d）播放中

图 6-2　录音控制演示效果

在 uniappch06 项目中，新建页面 recorder.vue，然后在 pages.json 中将 recorder.vue 页面作为第一项。

recorder.vue 代码如下。

```
<template>
    <view>
        <view >
            <!-- 初始页面，没有录音、播放操作 -->
            <block v-if="!recording && !playing && !hasRecord">
                <view class="page-body-time">
                    <text class="time-big">{{formatedRecordTime}}</text>
                </view>
                <view class="page-body-buttons">
                    <view class="page-body-button"></view>
                    <view class="page-body-button" @click="startRecord">
                        <image src="../../static/record.png"></image>
                    </view>
                    <view class="page-body-button"></view>
                </view>
            </block>
            <!-- 正在录音页面 -->
            <block v-if="recording === true">
                <view class="page-body-time">
                    <text class="time-big">{{formatedRecordTime}}</text>
                </view>
                <view class="page-body-buttons">
                    <view class="page-body-button"></view>
                    <view class="page-body-button" @click="stopRecord">
                        <view class="button-stop-record"></view>
                    </view>
                    <view class="page-body-button"></view>
                </view>
            </block>
            <!-- 录音完成，此时录音还未处于播放状态 -->
            <block v-if="hasRecord === true && playing === false">
                <view class="page-body-time">
                    <text class="time-big">{{formatedPlayTime}}</text>
```

```html
                <text class="time-small">{{formatedRecordTime}}</text>
            </view>
            <view class="page-body-buttons">
                <view class="page-body-button" @click="playVoice">
                    <image src="../../static/play2.png"></image>
                </view>
                <view class="page-body-button" @click="clear">
                    <image src="../../static/trash.png"></image>
                </view>
            </view>
        </block>
        <!-- 已完成录音,此时录音处于播放状态 -->
        <block v-if="hasRecord === true && playing === true">
            <view class="page-body-time">
                <text class="time-big">{{formatedPlayTime}}</text>
                <text class="time-small">{{formatedRecordTime}}</text>
            </view>
            <view class="page-body-buttons">
                <view class="page-body-button" @click="stopVoice">
                    <image src="../../static/stop.png"></image>
                </view>
                <view class="page-body-button" @click="clear">
                    <image src="../../static/trash.png"></image>
                </view>
            </view>
        </block>
    </view>
</view>
</template>
<script>
    import * as util from '../../common/util.js' //导入util.js,并取别名为util
    var playTimeInterval = null; //播放录音时的计时器
    var recordTimeInterval = null; //录音时的计时器
    var recorderManager = null; //recorderManager 对象
    var music = null; //innerAudioContext 对象
    export default {
        data() {
            return {
                recording:false, //录音中
                playing:false, //播放中
                hasRecord:false, //是否有了一个录音记录
                tempFilePath:'',
                recordTime:0,
                playTime:0,
                formatedRecordTime:'00:00:00', //录音的总时间
                formatedPlayTime:'00:00:00' //播放录音的当前时间
            }
        },
        onUnload:function() {
            this.end();
        },
```

```
onLoad:function() {
    music = uni.createInnerAudioContext();
    music.onEnded(() => {
        clearInterval(playTimeInterval)
        var playTime = 0
        console.log('play voice finished')
        this.playing = false;
        this.formatedPlayTime = util.formatTime(playTime);
        this.playTime = playTime;
    });
    recorderManager = uni.getRecorderManager();
    recorderManager.onStart(() => {
        console.log('recorder start');
        this.recording = true;
        recordTimeInterval = setInterval(() => {
            this.recordTime += 1;
            this.formatedRecordTime = util.formatTime(this.recordTime);
        }, 1000)
    });
    recorderManager.onStop((res) => {
        console.log('on stop');
        music.src = res.tempFilePath;
        console.log(music.src);

        this.hasRecord = true;
        this.recording = false;
    });
    recorderManager.onError(() => {
        console.log('recorder onError');
    });
},
methods:{
    async startRecord() { //开始录音
        recorderManager.start({
            format:'mp3'
        });
    },
    stopRecord() { //停止录音
        recorderManager.stop();
        clearInterval(recordTimeInterval);
    },
    playVoice() {
        console.log('play voice');
        this.playing = true;
        playTimeInterval = setInterval(() => {
            this.playTime += 1;
            this.formatedPlayTime = util.formatTime(this.playTime);
        }, 1000)
        music.play();
    },
    stopVoice() {
        clearInterval(playTimeInterval)
        this.playing = false;
```

```
                    this.formatedPlayTime = util.formatTime(0);
                    this.playTime = 0;
                    music.stop();
                },
                end() {
                    music.stop();
                    recorderManager.stop();
                    clearInterval(recordTimeInterval)
                    clearInterval(playTimeInterval);
                    this.recording = false, this.playing = false, this.hasRecord = false;
                    this.playTime = 0, this.recordTime = 0;
                    this.formatedRecordTime = "00:00:00", this.formatedRecordTime = "00:00:00";
                },
                clear() {
                    this.end();
                }
            }
        }
</script>
<style>
    image {
        width:150rpx;
        height:150rpx;
    }
    .page-body-wrapper {
        justify-content:space-between;
        flex-grow:1;
        margin-bottom:300rpx;
    }
    .page-body-time {
        display:flex;
        flex-direction:column;
        align-items:center;
    }
    .time-big {
        font-size:60rpx;
        margin:20rpx;
    }
    .time-small {
        font-size:30rpx;
    }
    .page-body-buttons {
        margin-top:60rpx;
        display:flex;
        justify-content:space-around;
    }
    .page-body-button {
        width:250rpx;
        text-align:center;
    }
    .button-stop-record {
        width:110rpx;
        height:110rpx;
```

```
        border:20rpx solid #fff;
        background-color:#f55c23;
        border-radius:130rpx;
        margin:0 auto;
    }
</style>
```

6.1.3 图片控制

uni-app 为开发者提供了一系列的图片操作 API，包括 uni.chooseImage、uni.previewImage、uni.getImageInfo、uni.saveImageToPhotosAlbum、uni.compressImage 等。

1. uni.chooseImage(OBJECT)

通过 uni.chooseImage(OBJECT)可以从本地相册中选择图片或使用相机拍照来选择图片。App 平台如需要更丰富的相机拍照 API（如直接调用前置摄像头的 API），可参考 H5 的扩展规范 plus.camera。从微信小程序基础库 2.21.0 开始，uni.chooseImage(OBJECT) 停止维护，请使用 uni.chooseMedia(OBJECT) 代替它。

uni.chooseImage(OBJECT)的参数说明如表 6-6 所示。

表 6-6　uni.chooseImage(OBJECT)的参数说明

参数	类型	必填	说明	平台差异
count	Number	否	最多可以选择的图片张数，默认值为 9	—
sizeType	Array<String>	否	original 表示原图，compressed 表示压缩图，默认二者都有	App、微信小程序、支付宝小程序、百度小程序支持
extension	Array<String>	否	根据文件扩展名过滤，每一项都不能是空字符串。默认不过滤	H5（HBuilderX 2.9.9+）支持
sourceType	Array<String>	否	album 表示从相册选择图片，camera 表示使用相机拍照，默认二者都有。如需直接使用相机或直接从相册选图片，请只使用其中一个	
crop	Object	否	图像裁剪参数，设置后 sizeType 失效	App 3.1.19+支持
success	Function	是	调用成功则返回图片的本地文件路径列表	
fail	Function	否	接口调用失败的回调函数	小程序、App 支持
complete	Function	否	接口调用结束的回调函数（调用成功、失败都会执行）	

crop 为 Object 类型，其参数说明如表 6-7 所示。

表 6-7　crop 的参数说明

参数名	类型	必填	说明
quality	Number	否	取值范围为 1~100，数值越小，图片质量越低（仅对 JPG 格式有效）。默认值为 80
width	Number	是	裁剪的宽度，单位为 px，用于计算裁剪宽高比

续表

参数名	类型	必填	说明
height	Number	是	裁剪的高度，单位为 px，用于计算裁剪宽高比
resize	Boolean	否	是否将 width 和 height 作为裁剪保存图片的真实像素值。默认值为 true。注意，它设置为 false 时在裁剪编辑界面显示图片的像素值，设置为 true 时不显示

success 的说明如下。

tempFilePaths：图片的本地文件路径列表，类型为 Array<String>。

tempFiles：图片的本地文件列表，其中的每一项都是一个 File 对象，类型为 Array<File>。File 对象的参数说明如表 6-8 所示。

表 6-8 File 对象的参数说明

参数	类型	说明
path	String	本地文件路径
size	Number	本地文件大小，单位为 B
name	String	包含扩展名的文件名称，仅 H5 支持
type	String	文件类型，仅 H5 支持

示例代码如下。

```
uni.chooseImage({
    count:6, //默认值为 9
    sizeType:['original', 'compressed'], //可以指定是原图还是压缩图，默认二者都有
    sourceType:['album'], //从相册中选择图片
    success:function (res) {
        console.log(JSON.stringify(res.tempFilePaths));
    }
});
```

2. uni.previewImage(OBJECT)

uni.previewImage(OBJECT)可以用来预览多张图片，并设置默认显示的图片，参数说明如表 6-9 所示。

表 6-9 uni.previewImage(OBJECT)的参数说明

参数	类型	必填	说明	平台差异
current	String	否	当前显示图片的链接，不填则默认为 urls 中的第一张的，注意，在 App1.95~1.98 中必填	
urls	Array<String>	是	需要预览的图片链接列表	
indicator	String	否	图片指示器样式，可取值：default 表示底部圆点指示器；number 表示顶部数字指示器；none 表示不显示指示器	App 支持
loop	Boolean	否	是否可循环预览，默认值为 false	App 支持

续表

参数	类型	必填	说明	平台差异
longPressActions	Object	否	长按图片显示操作菜单，如不填，就默认为保存相册	App 1.9.5+ 支持
success	Function	否	接口调用成功的回调函数	
fail	Function	否	接口调用失败的回调函数	
complete	Function	否	接口调用结束的回调函数（调用成功、失败都会执行）	

示例代码如下。

```
uni.chooseImage({
    count:6,
    sizeType:['original', 'compressed'],
    sourceType:['album'],
    success:function(res) {
        // 预览图片
        uni.previewImage({
            urls:res.tempFilePaths,
            longPressActions:{
                itemList:['发送给朋友', '保存图片', '收藏'],
                success:function(data) {
                    console.log('选中了第' + (data.tapIndex + 1) + '个按钮,第' + (data.index + 1) + '张图片');
                },
                fail:function(err) {
                    console.log(err.errMsg);
                }
            }
        });
    }
});
```

3. uni.getImageInfo(OBJECT)

uni.getImageInfo(OBJECT)用来获取图片信息，包括图片的宽度、图片的高度及图片返回的路径，参数说明如表 6-10 所示。对于网络图片，需先配置下载域名，uni.getImageInfo(OBJECT) 才能生效。

表 6-10 uni.getImageInfo(OBJECT)的参数说明

参数	类型	必填	说明
src	String	是	图片的路径，可以是相对路径、临时文件路径、存储文件路径、网络图片路径
success	Function	否	接口调用成功的回调函数
fail	Function	否	接口调用失败的回调函数
complete	Function	否	接口调用结束的回调函数（调用成功、失败都会执行）

success 返回参数说明如表 6-11 所示。

表 6-11 success 返回参数说明

参数	类型	说明
width	Number	图片宽度，单位为 px
height	Number	图片高度，单位为 px
path	String	返回图片的本地路径
orientation	String	拍照时设备的方向
type	String	图片格式

示例代码如下。

```
uni.chooseImage({
    count:1,
    sourceType:['album'],
    success:function (res) {
        uni.getImageInfo({
            src:res.tempFilePaths[0],
            success:function (image) {
                console.log(image.width);
                console.log(image.height);
            }
        });
    }});
```

4. uni.saveImageToPhotosAlbum(OBJECT)

uni.saveImageToPhotosAlbum(OBJECT)用于将图片保存到系统相册里，但需要用户授权，其参数说明如表 6-12 所示。

表 6-12 uni.saveImageToPhotosAlbum(OBJECT)的参数说明

参数	类型	必填	说明
filePath	Array\<String\>	是	图片文件路径，可以是临时文件路径，也可以是永久文件路径
success	Function	否	接口调用成功的回调函数
fail	Function	否	接口调用失败的回调函数
complete	Function	否	接口调用结束的回调函数（调用成功、失败都会执行）

uni.saveImageToPhotosAlbum(OBJECT)调用成功后会返回调用结果。

示例代码如下。

```
uni.chooseImage({
    count:1,
    sourceType:['camera'],
    success:function (res) {
        uni.saveImageToPhotosAlbum({
            filePath:res.tempFilePaths[0],
            success:function () {
                console.log('save success');
            }
        });
    }
});
```

5. uni.compressImage(OBJECT)

uni.compressImage(OBJECT)用于压缩图片。微信小程序支持将图片进行压缩，图片压缩质量可以根据自己的需求通过 quality 属性进行设置，压缩质量的范围为 0~100，数值越小，压缩质量越低，压缩率越高（仅对 JPG 图片有效）。uni.compressImage(OBJECT)的参数说明如表 6-13 所示。

表 6-13　uni.compressImage(OBJECT)的参数说明

参数	类型	必填	说明
src	String	是	图片的路径，可以是相对路径、临时文件路径、存储文件路径
quality	Number	否	图片压缩质量，范围为 0~100（默认值为 80），数值越小，压缩质量越低，压缩率越高（仅对 JPG 有效）
success	Function	否	接口调用成功的回调函数
fail	Function	否	接口调用失败的回调函数
complete	Function	否	接口调用结束的回调函数（调用成功、失败都会执行）

uni.compressImage(OBJECT)调用成功后会返回压缩成功的图片的临时路径 tempFilePath，微信开发者工具暂时不支持 wx.compressImage 调试，需要使用真机进行开发调试。

示例代码如下。

```
uni.compressImage({
  src:'/static/logo.jpg',
  quality:80,
  success:res => {
    console.log(res.tempFilePath)
  }
})
```

6.1.4　video 组件控制

视频 API 与 6.1.3 小节讲解的图片 API 类似，这里仅列举其相关方法。

- uni.chooseVideo(OBJECT)：用于拍摄视频或从手机相册中选择视频，返回视频的临时文件路径。
- uni.chooseMedia(OBJECT)：用于拍摄或从手机相册中选择图片或视频。
- uni.saveVideoToPhotosAlbum(OBJECT)：用于保存视频到系统相册。
- uni.getVideoInfo(OBJECT)：用于获取视频详细信息。
- uni.compressVideo(OBJECT)：用于压缩视频，开发者可通过 quality 属性指定压缩质量。
- uni.openVideoEditor(OBJECT)：用于打开视频编辑器，仅微信 2.120 以上版本支持。

本节主要讲解视频 API：uni.createVideoContext(videoId, this)。

uni.createVideoContext(videoId, this)用于创建并返回视频上下文对象 videoContext。在自定义组件中，第二个参数用于传入组件实例，以操作组件内的 video 组件。videoContext 对象的方法如表 6-14 所示。

表 6-14 videoContext 对象的方法

方法	参数	说明	平台差异
play	无	播放视频	
pause	无	暂停播放视频	
seek	position	跳转到指定位置，单位为 s	
stop		停止播放视频	微信小程序支持
sendDanmu	danmu	发送弹幕，danmu 包含两个属性，即 text、color	
playbackRate	rate	设置倍速播放，支持的倍速有 0.5、0.8、1.0、1.25、1.5。从微信基础库 2.6.3 起支持 2.0 倍速	
requestFullScreen	无	进入全屏模式，可传入{direction}参数，详见 video 组件	H5和字节跳动小程序不支持{direction}参数
exitFullScreen	无	退出全屏模式	
showStatusBar	无	显示状态栏，仅在 iOS 全屏模式下有效	微信、百度、QQ 小程序支持
hideStatusBar	无	隐藏状态栏，仅在 iOS 全屏模式下有效	微信、百度、QQ 小程序支持

【实例 6-3】视频及弹幕演示。

在 uniappch06 项目中，新建页面 video.vue。在该页面中演示调用 videoContext 对象的 play、pause、seek 方法，对 video 组件进行控制；调用 videoContext 对象的 sendDanmu 方法，发送弹幕。初始弹幕在一个数组中，与 video 组件关联。页面效果如图 6-3 所示。

实例 6-3

图 6-3 视频及弹幕演示效果

具体代码如下。

```
<template>
    <view>
        <video id="myVideo" :src="videoUrl" controls style="width:100%;height:
```

```
500rpx;" object-fit="fill" @timeupdate="getTime" :danmu-list="danmuList" enable-
danmu :danmu-btn="true"></video>
            当前播放时间:{{currentTime}}/{{duration}}
            <view class="bullet">
                <input type="text" placeholder="请输入弹幕内容" v-model="danmuValue"
class="input-bullet">
                <button @click="sendDanmu()" type="primary">发送弹幕</button>
            </view>
            <view class="btns">
                <button @click="play()">播放</button>
                <button @click="pause()">暂停</button>
                <button @click="goTime()">前进</button>
                <button @click="backTime()">后退</button>
            </view>

        </view>
    </template>
    <script>
        import * as util from '../../common/util.js' //导入util.js
        export default {
            data(){
                return { //《我和我的祖国》的视频地址,来源:网易云音乐
                    videoUrl:"……", //此处省略的是《我和我的祖国》的视频地址,地址见源代码
                    currentTime:"00:00",
                    duration:"00:00",
                    danmuList:[ {
                        text:'第 1s 出现的弹幕',
                        color:'#ff0000',
                        time:1
                    }, {
                            text:'第 3s 出现的弹幕',
                            color:'#ff00ff',
                            time:3
                    } ],
                    danmuValue:"",
                    currentTime:"",
                    currentTimeNum:0
                }
            },
            onReady(){ //创建 video 组件控制实例
                this.videoContext = uni.createVideoContext('myVideo')
            },
            methods:{
                play(){
                    this.videoContext.play();
                },
                pause(){
                    this.videoContext.pause();
                },
                //通过 video 组件的@timeupdate 事件获取视频播放时长和总时长
                getTime(e){
                    this.currentTimeNum=Math.floor(e.detail.currentTime);
```

```
                    this.currentTime=util.formatTime(Math.floor(e.detail.
currentTime));
                    if (this.currentTime.startsWith('00:')){    //当时长不足1h时，从
"hh:mm:ss"中去掉"hh:"
                        this.currentTime = this.currentTime.substring(3);
                    }
                    this.duration=util.formatTime(Math.floor(e.detail.duration));
                    if (this.duration.startsWith('00:')){
                        this.duration = this.duration.substring(3);
                    }
                },
                goTime(){  //前进
                    this.videoContext.seek(this.currentTimeNum+15);
                },
                backTime(){  //后退
                    this.videoContext.seek(this.currentTimeNum-15);
                },
                sendDanmu(){   //发送弹幕，弹幕内容为文本框中的内容，颜色为随机值
                    this.videoContext.sendDanmu({
                        text:this.danmuValue,
                        color:this.getRandomColor()
                    })
                },
                getRandomColor() {
                    const rgb = []
                    for (let i = 0; i < 3; ++i) {
                        let color = Math.floor(Math.random() * 256).toString(16)
                        color = color.length == 1 ? '0' + color :color
                        rgb.push(color)
                    }
                    console.log('#' + rgb.join(''));
                    return '#' + rgb.join('')
                }
            }
        }
    </script>
    <!-- 这里的样式使用 scss 设置，因此可以使用 uni.scss 中定义的变量。以下语句为 scss 格式
的，可以嵌套-->
    <style scoped lang="scss">
    .bullet {
      display:flex;
      align-items:center;
      margin:15px;
      .input-bullet{
            flex:1;
            margin-right:10px;
            border:1px solid $uni-color-success;
            padding:10px; }
        }
    .btns{
      display:flex;
      justify-content:space-evenly;
    }
    </style>
```

6.1.5　camera 组件控制

可以使用 uni.createCameraContext() 创建并返回相机上下文对象 CameraContext，CameraContext 与页面内唯一的 camera 组件绑定，操作对应的 camera 组件。camera 组件说明参见 4.4.1 小节。

CameraContext 对象的方法如表 6-15 所示。

表 6-15　CameraContext 对象的方法

方法	参数类型	说明	平台差异
takePhoto	Object	拍照，可指定成像质量，若调用成功则返回图片路径	
setZoom	Object	设置缩放级别	微信小程序 2.10.0+ 支持，京东小程序不支持
startRecord	Object	开始录像	京东小程序不支持
stopRecord	Object	结束录像，若调用成功则返回封面与视频	京东小程序不支持

以上方法的参数均为 Object 对象。下面对 Object 对象的参数进行说明。

takePhoto 方法的 Object 的参数有：quality、success、fail、complete。

setZoom 方法的 Object 的参数有：zoom、success、fail、complete。

startRecord 方法的 Object 的参数有：timeoutCallback、success、fail、complete。

stopRecord 方法的 Object 的参数有：compressed、success、fail、complete。

success、fail、complete 的类型均为回调函数，其中 takePhoto 方法的 success 会返回照片文件的临时路径 res={tempImagePath}，stopRecord 方法的 Success 会返回视频临时路径 res={tempThumbPath, tempVideoPaht}。

takePhoto 方法中 Object 的参数 quality 为 String 类型，表示成像质量，其值为 high（高质量）、normal（普通质量）、low（低质量）。

setZoom 方法中 Object 的参数 zoom 为 String 类型，表示缩放级别，为必填项。取值范围为[1, maxZoom]。maxZoom 可从@initdone 返回值中获取。zoom 可取小数，精确到小数后 1 位。

startRecord 方法中 Object 的参数 timeoutCallback 为回调函数，表示在超过 30s 或页面隐藏时会结束录像。

stopRecord 方法中 Object 的参数 compressed 为 Boolean 类型，表示启动视频压缩，仅微信小程序 2.10.0+支持。

【实例 6-4】演示相机拍照功能。

在本例中实现拍照后在界面中显示预览图片，点击图片可以查看图片，点击【保存图片】按钮则保存图片至相册，在 H5 平台下，可选择保存路径。运行效果如图 6-4 所示。

图 6-4 相机拍照运行效果

在 uniappch06 项目中新建页面 camera.vue，具体代码如下。

```
<template>
    <view>
        <camera device-position="back" flash="off" @error="error" style="width:100%;height:200px;"></camera>
        <view class="btnwrap">
            <button type="primary" @click="takePhoto">拍    照</button>
            <button type="primary" @click="savePhoto">保  存  图  片</button>
        </view>
        <view class="txt">预览图片——点击放大</view>
        <image mode="aspectFit" :src="src" :data-src="src" @tap="previewImage"></image>
    </view>
</template>
<script>
export default {
    data() {
        return {
            src:"",
            imageList:[]
        }
    },
    methods:{
        takePhoto() {
            const ctx = uni.createCameraContext();
            ctx.takePhoto({
                quality:'high',
```

```
            success:(res) => {
                this.src = res.tempImagePath
            }
        });
    },
    previewImage(e){
        var current = e.target.dataset.src;
        this.imageList = this.imageList.concat(current);
        uni.previewImage({
            current:current,
            urls:this.imageList
        })
    },
    savePhoto(){
        uni.saveImageToPhotosAlbum({
            filePath:this.src,
             success:function(){
                 console.log('保存成功！');
             }
        })
    },
     error(e) {
        console.log(e.detail);
     }
  }
}
</script>
<style>
.btnwrap{
    display:flex;
    justify-content:center;
    margin:5px;
  }
.txt { margin:5px; }
</style>
```

6.2 文件操作

uni-app 为开发者提供了一系列的文件操作 API，包括文件保存、得到文件信息、打开文档、删除本地文件等。本节讲解与文件保存和文件选择操作相关的 API。

6.2.1 文件保存

在媒体控制中，通过拍照、录音、录像等获得的文件的临时路径，在应用本次启动期间可以正常使用，如需实现永久访问，需主动调用 uni.saveFile(OBJECT)方法。或者调用 uni.saveImageToPhotosAlbum(OBJECT)保存图片到系统相册，或者调用 uni.saveVideoToPhotosAlbum(OBJECT)保存视频到系统相册。这 3 个方法的参数说明如表 6-16 所示。

表 6-16　3 个方法的参数说明

方法	参数	success 返回参数值	平台差异
uni.saveFile	tempFilePath、success、fail、complete	savedFilePath	H5、快手小程序不支持
uni.saveImageToPhotosAlbum	filePath、success、fail、complete	path、errMsg	H5 不支持
uni.saveVideoToPhotosAlbum	filePath、success、fail、complete	errMsg	H5、京东小程序不支持

补充说明如下。

tempFilePath：临时文件路径。

filePath：图片文件路径，可以是临时文件路径也可以是永久文件路径，不支持网络图片路径。

uni.saveImageToPhotosAlbum 中的 success 返回参数值 path，仅 App 支持。

6.2.2　文件选择

在文件上传、图片上传或查看图片、视频时都会选择文件。uni-app 提供了多个选择文件 API，根据不同的应用或平台，可以选择对应的 API。

（1）uni.chooseFile(OBJECT)：选择文件，仅 H5 支持。

（2）uni.chooseImage(OBJECT)：从本地相册选择图片或使用相机拍照。

（3）uni.chooseVideo(OBJECT)：拍摄视频或从手机相册中选择视频，返回视频的临时文件路径。

（4）uni.chooseMedia(OBJECT)：拍摄或从手机相册中选择图片或视频，微信小程序 2.10.0+、字节跳动小程序、飞书小程序支持。

示例代码如下。

```
uni.chooseImage({
    count:1,
    sourceType:['camera'],
    success:function (res) {
        uni.saveImageToPhotosAlbum({
            filePath:res.tempFilePaths[0],
            success:function () {
                console.log('save success');
            }
        });
    }
});
```

6.3　设备操作

许多时候我们需要获得设备的品牌、型号、操作系统版本、地理位置等信息，uni-app 提供一系列的 API 用于获得与设备、系统相关的信息。

6.3.1 系统 API

uni-app 提供了得到系统信息、设备信息、窗体信息、设备设置等的多个 API。这里讲解 uni.getSystemInfo、uni.getDeviceInfo，其他 API 请读者参考 uni-app 官网。

1. uni.getSystemInfo(OBJECT)

uni.getSystemInfo(OBJECT)用于异步获取系统信息，包括设备 ID、设备类型、操作系统的名称及版本、屏幕宽度及高度等信息。其参数说明如表 6-17 所示。

表 6-17 uni.getSystemInfo(OBJECT)的参数说明

参数	类型	必填	说明
success	Function	是	接口调用成功的回调函数
fail	Function	否	接口调用失败的回调函数
complete	Function	否	接口调用结束的回调函数（调用成功、失败都会执行）

【实例 6-5】新建一个项目 uniappSys，在 index.vue 页面中添加如下代码。

```
onLoad() {
    uni.getSystemInfo({
        success:function(res){
            console.log("系统信息:",res);
        }
    })
}
```

实例 6-5

在微信开发者工具中可以得到如下输出内容（截取了部分内容），如图 6-5 所示。

```
系统信息 ▼{errMsg: "getSystemInfo:ok", model: "iPhone 6/7/8", pixelRatio: 2, windowWidth: 375, windowHeight: 603, …}
    SDKVersion: "2.27.0"
    appId: "__UNI__7589783"
    appLanguage: "zh-Hans"
    appName: "uniappSys"
    appVersion: "1.0.0"
    appVersionCode: "100"
    batteryLevel: 100
    benchmarkLevel: 1
    bluetoothEnabled: true
    brand: "devtools"
    browserName: undefined
    browserVersion: undefined
    cameraAuthorized: true
    deviceBrand: "devtools"
    deviceId: "16667637460915932646"
```

图 6-5 输出的系统信息

2. uni.getDeviceInfo()

uni.getDeviceInfo()用于获取设备基础信息，返回的内容有设备品牌、设备 ID、设备型号、设备类型、设备方向、操作系统及版本等信息。

示例代码如下。

```
console.log("设备信息:" , uni.getDeviceInfo());
```

在微信开发者工具中可以得到如下输出内容（部分），如图 6-6 所示。

图 6-6 输出的设备信息（部分）

6.3.2 扫码

uni.scanCode(OBJECT)用于打开客户端扫码界面，扫码成功后返回对应的结果。其参数说明如表 6-18 所示。

表 6-18 uni.scanCode(OBJECT)的参数说明

参数	类型	必填	说明	平台差异
onlyFromCamera	Boolean	否	是否只能通过相机扫码，不允许从相册选择图片	字节跳动、百度、支付宝小程序不支持
scanType	Array	否	扫码类型，如 barCode（表示一维码）、qrCode（表示二维码）、datamatrix、pdf417	字节跳动小程序不支持
success	Function	否	接口调用成功的回调函数	
fail	Function	否	接口调用失败的回调函数（识别失败、用户取消等情况下触发）	
complete	Function	否	接口调用结束的回调函数	

success 返回参数说明如表 6-19 所示。

表 6-19 success 返回参数说明

参数	说明	平台差异
result	扫码的内容	
scanType	扫码的类型	App、微信、百度、QQ、京东、支付宝小程序支持
charSet	扫码的字符集	App、微信、百度、QQ、京东小程序支持
path	当扫描当前应用的合法二维码时，会返回此字段，其值为二维码携带的 path	微信、QQ、京东小程序支持
rawData	原始数据，使用 base64 编码	微信、QQ、京东、支付宝小程序支持

【实例 6-6】扫码演示。

在 uniappSys 的 index.vue 页面中，设置 3 个按钮，它们分别用于调用相机扫码、在相册中选择图片、条形码扫码。因为需要访问相册和相机，所以将进行真机测试。在使用微信开发者工具进行演示时，会打开文件选择器选择图片进行扫码。具体代码如下。

实例 6-6

```
<template>
    <view class="content">
```

```html
            <view class="btns" >
                <button type="default" @click="scanFromCamera" class="scanbtn">相机扫码</button>
                <button type="primary" @click="scanFromBoth" class="scanbtn">相册扫码</button>
                <button type="default" @click="scanFromBar" class="scanbtn">条形码扫码</button>
            </view>
            <view class="result">
                <text>条形码类型:{{scanType}}</text> </br>
                <text>条形码内容:{{result}}</text>
            </view>
        </view>
    </template>
    <script>
        export default {
            data() {
                return {
                    scanType:'',
                    result:''
                }
            },
            methods:{
                scanFromCamera(){// 只允许通过相机扫码
                    uni.scanCode({
                        onlyFromCamera:true,
                        success:(res) =>   {
                            this.scanType =  res.scanType;
                            this.result= res.result;
                        }
                    });
                },
                scanFromBoth(){ // 允许通过相机和相册扫码
                    uni.scanCode({
                        success:(res) => {
                            this.scanType =  res.scanType;
                            this.result= res.result;
                        }
                    });
                },
                scanFromBar(){   // 扫描条形码
                    uni.scanCode({
                        scanType:['barCode'],
                        success:(res) =>   {
                            this.scanType =  res.scanType;
                            this.result= res.result;
                        }
                    });
                }
            }
        }
    </script>
```

```
<style lang="scss">
    .content {
        display:flex;
        flex-direction:column;
        align-items:center;
        justify-content:center;
    }
    .btns{
        width:100%;
        display:flex;
        flex-direction:row;
        justify-content:space-between;
        margin:10px;
        .scanbtn{
            margin-bottom:10px;
        }
    }
    .result {    width:100%;    }
</style>
```

在微信开发者工具中,点击【真机调试】按钮,选择【二维码真机调试】生成微信二维码,如图6-7(a)所示。使用微信扫码进行测试,测试效果如图6-7(b)所示。

(a)生成微信二维码

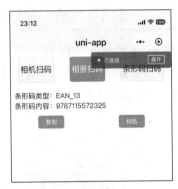

(b)测试效果

图 6-7　扫码运行效果

6.3.3　拨打电话

uni.makePhoneCall(OBJECT)用于拨打电话,其参数说明如表 6-20 所示。

表 6-20　uni.makePhoneCall(OBJECT)的参数说明

参数	类型	必填	说明
phoneNumber	String	是	需要拨打的电话号码
success	Function	否	接口调用成功的回调函数
fail	Function	否	接口调用失败的回调函数
complete	Function	否	接口调用结束的回调函数(调用成功、失败都会执行)

示例代码如下。

```
uni.makePhoneCall({
    phoneNumber:'114'  //仅为示例
});
```

6.3.4 剪贴板

uni.setClipboardData(OBJECT)：用于设置剪贴板的内容。其参数说明如表 6-21 所示。

uni.getClipboardData(OBJECT)：用于获取剪贴板的内容。其参数有 success、fail、complete 等，返回的值中有 res.data，为剪贴板中的内容。

表 6-21 uni.setClipboardData(OBJECT)的参数说明

参数	类型	必填	说明	平台差异
data	String	是	需要设置的内容	
showToast	Boolean	否	是否弹出提示框，默认弹出提示框	App、H5（3.2.13+）支持
success	Function	否	接口调用成功的回调函数	
fail	Function	否	接口调用失败的回调函数	
complete	Function	否	接口调用结束的回调函数	

【实例 6-7】剪贴板的演示。

在实例 6-6 的基础上添加【复制】【粘贴】按钮，点击【复制】按钮将扫码结果复制到剪贴板中，点击【粘贴】按钮弹出提示框，并在提示框中显示剪贴板的内容。演示效果如图 6-8 所示。

（a）点击【复制】按钮

（b）点击【粘贴】按钮

图 6-8 剪贴板演示效果

在<template></template>部分添加以下代码。

```
<view class="btns">
    <button @click="copy" type="primary" size="mini">复制</button>
    <button @click="paste" type="primary" size="mini">粘贴</button>
</view>
```

在<script></script>的 methods 中添加以下方法。

```
copy() {
    uni.setClipboardData({
        data:this.result,
        success:function() {
            console.log('success');
        }
    });
},
paste() {
    uni.getClipboardData({
        success:(res) => {
            uni.showToast({
                title:res.data
            })
        }
    })
}
```

6.4 登录

uni-app 提供一系列第三方服务 API，如第三方登录、第三方支付、分享等 API。登录功能是移动应用必不可少的功能，可以有多种登录方式，如手机验证码、扫码、用户密码、第三方账号等。使用第三方账号登录的有很多，如淘宝账号、微博账号、QQ 账号、微信账号登录等。以第三方账号登录时，过程一般比较类似。下面以微信小程序中微信账号登录为例讲解登录过程。微信小程序的登录可以简单地分为以下几个基本步骤。

（1）在微信小程序里使用 uni.login 方法获取用户登录凭证（code）。

（2）将 code 和 AppID、AppSecret（另外需在应用下申请打开微信登录）、grant_type 这 4 个参数发送到自己开发的后台服务器上，在后台服务器上请求路径 https://api.weixin.qq.com/sns/jscode2session（该地址是微信请求的接口地址，直接打开不能使用，需要传递以上参数才能使用）。同时传递这 4 个参数，就能获取唯一标识（openid）和会话密钥（session_key）。

（3）获取 openid 和 session_key 后，在自己开发的后台服务器上生成自己的 sessionId。

（4）微信小程序可以将服务器生成的 sessionId 保存到本地缓存（Storage）。

（5）后续用户进入微信小程序时，先从本地缓存中获取 sessionId，然后将 sessionId 传输到服务器上进行查询以维护登录状态。

下面一起来看这些步骤的实现。

1. 用 uni.login(OBJECT)获取 code

微信小程序使用 uni.login(OBJECT)来获取 code。用户被允许登录后，回调内容中有 code（有效时间为 5min）。

示例代码如下。

```
methods:{
wxlogin() {
   uni.login({
     provider:'weixin', //使用微信登录
     success:function(res){
       var code = res.code; //用户登录凭证
       if(code){
         console.log('获取用户登录凭证:'+code);
       }else{
         console.log('获取用户登录凭证失败');
       }
     }
   })
 }
})
```

2. 将 code 发送到开发者后台服务器获取 openid 和 session_key

开发者服务器提供一个后台接口，用来接收 code。

```
methods:{
wx.login() {
   uni.login({
     success:function(res){
       provider:'weixin', //使用微信登录
       var code = res.code; //用户登录凭证
       if(code){
         console.log('获取用户登录凭证:'+code);
         uni.request({ //请求后台服务器，传输用户登录凭证
           url:'https://www.my-domain.com/wx/onlogin',   // 示例地址，请用真实
                                                        // 地址代替
           data:{ code:code }
         })
       }else{        console.log('获取用户登录凭证失败');          }
     }
   })
 }
})
```

3. 开发者后台服务器使用 code 获取 openid 和 session_key

开发者后台服务器接收 code 后，将其与 AppID、AppSecret、grant_type 这 3 个参数一起请求微信服务器接口 https://api.weixin.qq.com/ sns/jscode2session，以获取 session_key 和 openid。其中，session_key 是对用户数据进行加密签名的密钥。为了应用安全，session_key 不应该在网络上传输。

登录接口地址为：https://api.weixin.qq.com/sns/jscode2session?appid=APPID&secret=SECRET& js_code=JSCODE&grant_type=authorization_code。

参数说明：appid 为小程序的 AppID；secret 为小程序的 App Secret；js_code 为登录时获取的 code；grant_type 设置为"authorization_code"固定值。

后台服务器请求微信服务器的代码可以使用 Java、PHP、C++、Node.js 等多种语言。

4. 开发者后台服务器生成自己的 sessionId

开发者后台服务器获取到 openid 和 session_key 后，需要生成自己的 sessionId。生成规则可以由自己制定，可以拼接成字符串，也可以拼接成字符串后再用 MD5 加密等。生成的 sessionId 需要在开发者后台服务器中保存起来，小程序进行在校验登录或者进行登录后才能做的操作时，都需要在开发者后台服务器中验证 sessionId。sessionId 可以保存到缓存 Memcached、Redis 或内存中。

5. 小程序客户端保存 sessionId

小程序客户端是没有类似于浏览器客户端的 cookies 或者 session 机制的，但是可以利用小程序的 Storage 缓存机制来保存 sessionId，在需要登录态才能发起请求的时候传递 sessionId，从而不用每次都重新登录。在之后调用那些登录后才有权限访问的后台服务时，可以将保存在 Storage 中的相应 sessionId 取出并携带在请求中，传递到后台服务，后台服务获取到该 sessionId 后，从 Redis 或者内存中查找是否有该 sessionId 的存在，若存在即确认该 sessionId 是有效的，继续执行后续的代码，否则进行错误处理。

6. 使用 uni.checkSession（OBJECT）检查登录态是否过期

微信小程序可以使用 uni.checkSession（OBJECT）来检查登录态是否过期，如果过期就重新登录。

通过 uni.login 获得的用户登录态具有一定的时效性。用户如果很久未使用小程序，其登录态有可能失效；如果用户一直在使用小程序，则其登录态将一直保持有效。具体时效逻辑由微信维护，对开发者透明。开发者只需要调用 uni.checkSession 检测当前用户登录态是否有效即可。

登录态过期后，开发者可以调用 uni.login 获取新的用户登录态。若调用成功，则说明当前 session_key 未过期；调用失败则说明 session_key 已过期。示例代码如下：

```
uni.checkSession({
    success:function(){    //session_key 未过期    },
    fail:function(){    //session_key 已过期
        uni.login()    }
})
```

6.5 案例：仿网易云音乐 App 的音乐播放器

本案例后端接口为 GitHub 网站上的开源接口，需本地部署服务器，具体参见本书电子资源。部署后 API 服务器地址为 http://localhost:3000，如图 6-9 所示。本案例仅供学习使用，不得商用。

本案例的运行效果如图 6-10 所示。其中图 6-10（a）所示为首页（榜单列表页），图 6-10（b）所示为榜单详情页，图 6-10（c）所示为播放歌曲页，播放歌曲时动态显示歌词，以及显示歌曲的前 5 条评论。本案例没有使用 uni-ui、uView 等组件，读者可以对照图 6-10 选用组件方式来实现。

案例：仿网易云音乐 App 的音乐播放器

图 6-9 部署 API 服务器

（a）首页

（b）榜单详情页

（c）播放歌曲页

图 6-10 播放器运行效果

实现步骤

1. 新建项目

新建一个 uni-app 项目 MusicCloudDemo，选择 Vue2，新建 list.vue、detail.vue 页面（自动生成一个 index.vue 页面）。修改 pages.json 文件，将这 3 个页面的 navigationStyle 属性设置为 custom。这样页面顶部的导航为自定义。

2. 准备字体图标和全局样式

在 iconfont 网站搜索图片"左箭头"、"首页"、"播放"（两个）、"暂停"、"分享"，然后将它们添加到购物车，添加完成后，点击购物车图标，选择【下载代码】。在项目根目录下新建一个 common 文件夹，将下载的压缩文件中的 iconfont.css、iconfont.ttf 文件复制到 common 中。这里采用字体图标的形式。

修改 iconfont.css 文件中的 iconfont.ttf 文件引用路径。

```
@font-face {
  font-family:"iconfont"; /* Project id    */
  src:url('~@/common/iconfont.ttf?t=1673014428887') format('truetype');
}
```

在 App.vue 中引用 iconfont.css 文件，并设置全局的样式。

```css
<style>
/*每个页面公共 CSS */
@import '@/common/iconfont.css';
.container {
    width:100%;
    height:calc(100vh - 75px);
    overflow:hidden;
}
.container scroll-view {
    height:100%;
}
.topbg {
    width:100%;
    height:100vh;
    position:fixed;
    background-size:cover;
    background-position:center 0;
    filter:blur(10px);
    transform:scale(1.2);
}
</style>
```

3. 自定义一个页面头部组件 musichead

在项目根目录下新建 components 文件夹，然后在 components 文件夹上单击鼠标右键，在弹出的菜单中选择【新建组件】，勾选【创建同名目录】，让组件符合"components/组件名称/组件名称.vue"的目录结构。这样就可以不引用、注册，直接在页面中使用组件。

musichead.vue 的代码如下。

```vue
<template>
    <view class="music-head" :style="{ color :color }">
        <view v-if="icon" class="music-head-icon" :class="{ 'music-head-iconBlack' :iconBlack }">
            <text @tap="gotoBack" class="iconfont icon-jzuojiantou"></text> |
            <text @tap="gotoHome"   class="iconfont icon-shouye"></text>
        </view>
        {{ pageTitle }}
    </view>
</template>
<script>
    export default {
        data() {
            return {
            };
        },
        props:['pageTitle', 'icon', 'color', 'iconBlack'],
        methods:{
            gotoBack() {
                uni.navigateBack();
            },
            gotoHome() {
                uni.navigateTo({
                    url:'/pages/index/index'
                });
```

```
            }
        }
    }
</script>
<style>
    .music-head {
        height:75px;
        text-align:center;
        line-height:75px;
        font-size:16px;
        font-weight:600;
        color:black;
        overflow:hidden;
        position:relative;
    }
    .music-head-icon {
        display:flex;
        justify-content:space-evenly;
        width:97px;
        height:31px;
        background:rgba(0, 0, 0, 0.3);
        border-radius:20px;
        position:absolute;
        left:10px;
        top:25px;
        line-height:31px;
        color:white;
    }
    .music-head-iconBlack {
        color:black;
        background:white;
        border:1px black solid;
    }
</style>
```

【代码解析】

musichead 组件为页面头部的导航栏，包括页面导航、页面标题。页面导航包含后退导航和返回首页导航。后退导航使用的是"左箭头"字体图标，点击后退导航，将调用 uni.navigateBack 方法。返回首页导航使用的是"首页"字体图标，点击返回首页导航，将调用 uni.navigateTo 方法。

该组件有 4 个属性：pageTitle、icon、color、iconBlack，具体含义如下。

➢ pageTitle：用于设置导航栏上的页面标题。

➢ icon：用于设置是否显示导航图标，值为 true 或 false。

➢ color：用于设置颜色值，这里用于修改标题的文字颜色。

➢ iconBlack：用于设置黑色的导航。设置该属性时，导航栏的导航部分变为白底、黑字、黑色边框。

示例代码如下。

```
<musichead pageTitle="网易云音乐榜单" :icon="false" color="#e0665b"></musichead>
```

4. 数据格式过滤

一般后端的日期用数字表示,用从 1970 年 1 月 1 日 0 点 0 分 0 秒开始到现在经过的毫秒数来表示日期。在页面中显示的日期格式一般为 "××××年××月××日"。对于比较大的数据,为了更加直观,往往显示为 "×.×亿" 或 "×.×万"。在 main.js 文件中,在创建 Vue 实例之前定义两个全局过滤器:formatTime、numberFormat。过滤器应该添加在 JavaScript 表达式的尾部,由管道符号(|)指示。例如:

```
<text>{{ 数字 | formatNumber}}</text>
```

main.js 的部分代码如下。

```
import App from './App'

Vue.filter('formatNumbert',function(value) {
    var param = {};
    var k = 10000,
        sizes = ['', '万', '亿', '万亿'],
        i;
    if(value < k){
        param.value =value
        param.unit=''
    }else{
        i = Math.floor(Math.log(value) / Math.log(k));
        param.value = ((value / Math.pow(k, i))).toFixed(2);
        param.unit = sizes[i];
    }
    return param.value+param.unit;
});

Vue.filter('formatTime',function(value){
    var date = new Date(value);
    return date.getFullYear() + '年' + (date.getMonth() + 1) + '月' + date.getDate() + '日';
});
```

5. 首页 index.vue

首页用来显示网易云音乐的榜单列表。点击某一榜单,则进入对应的榜单详情页 list.vue,并传递榜单 "id"。通过 uni.request 来请求 http://localhost:3000/toplist/detail 地址得到榜单信息(一定要启动本地 API 服务)。

`<script></script>`模块的代码如下。

```
<script>
    export default {
        data() {
            return {    topList:[]      }
        },
        onLoad() {
            uni.request({
                url:'http://localhost:3000/toplist/detail',
                method:'GET',
                data:{},
                success:res => {
```

```
                    let result = res.data.list;
                    result.length = 10;
                    this.topList = result;
                    console.log(result);
                },
                fail:(err) => {
                    console.log(err);
                }
            });
        },
        methods:{
            handleToList(id) {
                uni.navigateTo({
                    url:'/pages/list/list?id=' + id
                });
            }
        }
    }
</script>
```

请求成功后返回的数据中 res.data.list 为榜单数组。将榜单数组赋给 topList，然后设置 topList 的长度为 10。在页面中输出前 10 个榜单数据。读者可以查看控制台，获取各字段及相关信息。部分字段如图 6-11 所示。

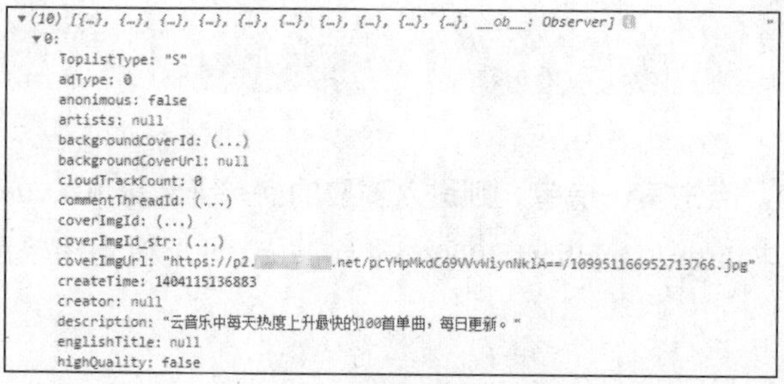

(a) 字段内容 1 　　　　　　　　　　(b) 字段内容 2

图 6-11　控制台输出信息

在首页中显示图片（coverImgUrl）、更新频率（updateFrequency）、歌曲（tracks）。因为只有前 3 个榜单的 tracks 有值，故如果 tracks 的长度为 0，则在榜单右边显示榜单名（name）和更新频率。

<template></template>模块的代码如下。

```
<template>
    <view class="index">
```

```
            <musichead pageTitle="网易云音乐榜单" :icon="false" color="#e0665b" >
</musichead>
            <view class="container">
                <scroll-view scroll-y="true">
                    <view class="index-list">
                        <view v-for="(item,index) in topList" :key="index" @tap=
"handleToList(item.id)">
                            <view v-if="index < 4" class="index-list-item">
                                <view class="index-list-img">
                                    <image :src="item.coverImgUrl" mode=""></image>
                                    <text>{{ item.updateFrequency }}</text>
                                    <span class="bi bi-play "></span>
                                </view>
                                <view class="index-list-text">
                                    <view v-for="(musicItem , index) in item.tracks"
:key="index">
                                        {{ index + 1 }}.{{musicItem.first}} - {{musicItem.second}}
</view>
                                </view>
                            </view>
                            <view v-else class="index-list-item">
                                <view class="index-list-img">
                                    <image :src="item.coverImgUrl" mode=""></image>
                                </view>
                                <view class="index-list-text">
                                    <view class="index-list-name">{{ item.name }}</view>
                                    <view>{{ item.updateFrequency }}</view>
                                </view>
                            </view>
                        </view>
                    </view>
                </scroll-view>
            </view>
    </template>
```

`<style></style>` 模块的代码如下。

```
    <style>
        .index-list {
            margin:0 40rpx;
        }
        .index-list-item {
            display:flex;
            margin-bottom:35rpx;
            background-color:aliceblue;
        }
        .index-list-img {
            width:212rpx;
            height:212rpx;
            margin-right:20rpx;
            border-radius:15rpx;
            overflow:hidden;
            position:relative;
        }
        .index-list-img image {
```

```
            width:100%;
            height:100%;
        }
        .index-list-img text {
            position:absolute;
            font-size:22rpx;
            color:white;
            bottom:15rpx;
            left:15rpx;
        }
        .index-list-text {
            flex:1;
            font-size:24rpx;
            line-height:50rpx;
            display:flex;
            flex-direction:column;
            justify-content:space-around;
        }
        .index-list-text .index-list-name {
            font-size:30rpx;
            font-weight:400;
            color:black;
        }
</style>
```

6. 榜单详情页 list.vue

根据首页传递的榜单"id",获取榜单详情。请求地址为:http://localhost:3000/playlist/detail?id='+id。同首页一样,在代码中输出请求成功后的值,查看相关字段。榜单详情页用到以下字段。

- 榜单信息:res.data.playlist。
- 歌曲列表:res.data.playlist.tracks。
- 榜单更新时间:res.data.playlist.updateTime。
- 榜单对应的图片:res.data.playlist.coverImgUrl。
- 榜单创建者的名字:res.data.playlist.creator.nickname。
- 榜单创建者的图标:res.data.playlist.creator.avatarUrl。
- 歌曲名称:res.data.playlist.tracks[i].al.name。
- 歌曲 ID:res.data.playlist.tracks[i].id。
- 歌曲对应图片:res.data.playlist.tracks[i].al.picUrl。
- 歌曲演唱者:res.data.playlist.tracks[i].ar[i].name。

<script></script>模块的代码如下。

```
<script>
    export default {
        data() {
            return {
                playlist:{},
                isShow:false,
                creator:''
```

```
            }
        },
        onLoad(options) {
            let id =options.id;
            uni.request({
                url:'http://localhost:3000/playlist/detail?id='+id,
                method:'GET',
                success:(res) => {
                    setTimeout(()=>{
                        console.log(res);
                        this.playlist = res.data.playlist;
                        this.isShow = true;
                        this.creator = this.playlist.creator;
                    },2000)
                },
                fail:(err) => {
                    console.log(err);
                },
                complete:() => {}
            })
        },
        methods:{
            gotoPlay(songid){
                uni.navigateTo({
                    url:"../detail/detail?songid=" + songid
                })
            }
        }
    }
</script>
```

榜单详情页中以列表的形式显示每首歌曲的图片、演唱者、歌曲名称，点击列表项将进入播放歌曲页 detail.vue，并传递歌曲"id"。

<template></template>模块的代码如下。

```
<template>
    <view class="list">
        <view class="topbg" :style="{backgroundImage:'url('+ playlist.coverImgUrl +')'}"></view>
        <musichead pageTitle="歌 单" :icon="true" color="white"></musichead>
        <view class="container">
            <scroll-view scroll-y="true">
                <view class="list-head">
                    <view class="list-head-img">
                        <image :src="playlist.coverImgUrl" mode=""></image>
                    </view>
                    <view class="list-head-text">
                        <view class="list-head-source">
                            <image :src="creator.avatarUrl" mode=""></image>
                            <text>{{ creator.nickname }}</text>
                        </view>
                        <view>{{ playlist.description }}</view>
                        <text>{{ playlist.updateTime | formatTime }}更新</text>
                    </view>
```

```html
                </view>
                <!-- #ifdef MP-WEIXIN -->
                <button v-show="isShow" class="list-share" open-type="share">
                    <text class="iconfont icon-share"></text>分享给微信好友
                </button>
                <!-- #endif -->
                <view class="list-music">
                    <view v-show="isShow" class="list-music-title">
                        <text class="iconfont icon-bofanganniu"></text>
                        <text>播放全部</text>
                        <text>(共{{ playlist.trackCount }}首)</text>
                    </view>
                    <view class="list-music-item" v-for="(item,index) in playlist.tracks" :key="item.id"
                        @tap="gotoPlay(item.id)">
                        <view class="list-music-top">{{ index + 1 }}</view>
                        <view class="list-music-image">
                            <image :src="item.al.picUrl" mode="widthFix"></image>
                        </view>
                        <view class="list-music-song">
                            <view>{{ item.name }}</view>
                            <view>
                                {{ item.ar[0].name }} - {{ item.name }}
                            </view>
                        </view>
                        <text class="iconfont icon-bofang1"></text>
                    </view>
                </view>
            </scroll-view>
        </view>
    </view>
</template>
```

<style></style>模块的代码如下。

```css
<style>
    .list-head {
        display:flex;
        margin:20rpx 30rpx;
        margin-top:0rpx;
    }
    .list-head-img {
        width:230rpx;
        height:230rpx;
        border-radius:15rpx;
        margin-right:40rpx;
        overflow:hidden;
        position:relative;
    }
    .list-head-img image {
        width:100%;
        height:100%;
    }
    .list-head-text {
        flex:1;
```

```css
    font-size:24rpx;
    color:#e9e9e9;
    display:flex;
    flex-direction:column;
    justify-content:space-between;
    padding-bottom:20rpx;
}
.list-head-source {
    width:100%;
    display:flex;
    align-items:center;
}
.list-head-text image {
    width:40rpx;
    height:40rpx;
    border-radius:50%;
    margin-right:15rpx;
}
.list-head-text view:nth-child(1) {
    font-size:34rpx;
    color:#ffffff;
}
.list-share {
    width:330rpx;
    height:72rpx;
    margin:0 auto;
    background:rgba(0, 0, 0, 0.4);
    text-align:center;
    line-height:72rpx;
    font-size:26rpx;
    color:white;
    border-radius:36rpx;
}

.list-share text {
    margin-right:15rpx;
}
.list-music {
    background:white;
    border-radius:50rpx;
    overflow:hidden;
    margin-top:45rpx;
}
.list-music-title {
    height:58rpx;
    line-height:58rpx;
    margin:30rpx 30rpx 30rpx 20rpx;
    display:flex;
    align-items:center;
}
.list-music-title text:nth-child(1) {
    font-size:40rpx;
}
.list-music-title text:nth-child(2) {
```

```css
            font-size:34rpx;
            margin:0 10rpx 0 25rpx;
        }
        .list-music-title text:nth-child(3) {
            font-size:28rpx;
            color:#b2b2b2;
        }
        .list-music-item {
            display:flex;
            margin:0 20rpx 50rpx 20rpx;
            align-items:center;
        }
        .list-music-top {
            width:56rpx;
            font-size:28rpx;
            color:#979797;
        }
        .list-music-image {
            margin-left:15rpx;
            margin-right:15rpx;
        }
        .list-music-image image {
            width:150rpx;
            height:150rpx;
        }
        .list-music-song {
            flex:1;
            line-height:40rpx;
        }
        .list-music-song view:nth-child(1) {
            font-size:30rpx;
            width:55vw;
            white-space:nowrap;
            overflow:hidden;
            text-overflow:ellipsis;
        }
        .list-music-song view:nth-child(2) {
            font-size:22rpx;
            color:#a2a2a2;
            width:55vw;
            white-space:nowrap;
            overflow:hidden;
            text-overflow:ellipsis;
        }
        .list-music-song image {
            width:34rpx;
            height:22rpx;
            margin-right:10rpx;
        }
        .list-music-item text {
            font-size:50rpx;
            color:#c8c8c8;
        }
</style>
```

7. 播放歌曲页 detail.vue

通过 list.vue 页面传递的歌曲"id"，查询歌曲信息、歌词信息、歌曲评论等。相关地址如下。

> 歌曲信息：http://localhost:3000/song/detail?id=歌曲 id。
> 歌曲：http://localhost:3000/song/url?id=歌曲 id。
> 歌曲评论：http://localhost:3000/comment/music?id=歌曲 id。
> 歌词：http://localhost:3000/lyric?id=歌曲 id。

（1）获取歌曲信息

将歌曲图片设置为页面背景和碟片中间的图片，将歌曲名字设置为页面标题，设置碟片中间的字体图标为暂停图标，效果如图 6-12 所示。

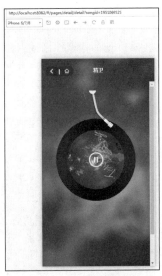

图 6-12 歌曲信息效果

完整的 detail.vue 的代码如下。

```
<template>
    <view class="detail">
        <view class="topbg" :style="{backgroundImage:'url('+ songDetail.al.picUrl +')'}"></view>
        <musichead :pageTitle="songDetail.name" :icon="true" color="white"></musichead>
        <view class="container">
            <scroll-view scroll-y="true">
                <view class="detail-play">
                    <image :src="songDetail.al.picUrl" mode=""></image>
                    <text class="iconfont" :class="playicon"></text>
                    <view></view>
                </view>
            </scroll-view>
        </view>
    </view>
</template>
<script>
    export default {
        data() {
            return {
                songid:'',
                songDetail:{
                    al:{   picUrl:''}
                },
                playicon:'icon-pause'
            }
        },
        onLoad(options) {
            this.songid = options.songid;
            this.playMusicDetail();
        },
        methods:{
```

```
            playMusicDetail() {
                uni.request({
                    url:'http://localhost:3000/song/detail?ids=' + this.songid,
                    method:'GET',
                    success:(res) => {
                        this.songDetail = res.data.songs[0];
                        console.log(this.songDetail);//在控制台输出歌曲信息,便于查看
                    }
                });
            },
        }
    }
</script>
<style>
    .detail-play {
        width:580rpx;
        height:580rpx;
        background:url(~@/static/disc.png);
        background-size:cover;
        margin:210rpx auto 44rpx auto;
        position:relative;
    }
    .detail-play image {
        width:380rpx;
        height:380rpx;
        border-radius:50%;
        position:absolute;
        left:0;
        top:0;
        right:0;
        bottom:0;
        margin:auto;
        animation:10s linear infinite move;
        animation-play-state:paused;
    }
    .detail-play text {
        width:100rpx;
        height:100rpx;
        font-size:100rpx;
        position:absolute;
        left:0;
        top:0;
        right:0;
        bottom:0;
        margin:auto;
        color:white;
    }
    .detail-play view {
        position:absolute;
        width:170rpx;
        height:266rpx;
        position:absolute;
        left:60rpx;
```

```
        right:0;
        margin:auto;
        top:-170rpx;
        background:url(~@/static/needle.png);
        background-size:cover;
    }
</style>
```

(2)实现播放歌曲功能,播放时碟片中的图片旋转

➢ 在 onLoad 函数中添加以下代码。

```
this.innerAudioContext = null;
```

➢ 在 methods 中添加播放歌曲的方法 playMusic。在该方法中,先创建一个内部音频上下文对象 innerAudioContext,然后请求地址 http://localhost:3000/song/url?id=id,得到歌曲的 URL,将 URL 赋给 innerAudioContext 对象的 src 属性,最后对 innerAudioContext 对象的 pause、play、stop 事件进行监听。具体代码如下。

```
playMusic(songId) {
        if (!this.innerAudioContext) {
            this.innerAudioContext = uni.createInnerAudioContext();
        }
        this.playicon = 'icon-bofang1';
        uni.request({
            url:'http://localhost:3000/song/url?id=' + this.songid,
            method:'GET',
            success:(res) => {
                console.log("url success");
                console.log(res.data.data[0].url);
                this.innerAudioContext.src = res.data.data[0].url;
                this.innerAudioContext.onPlay(() => {
                    this.playicon = 'icon-pause';      });
                this.innerAudioContext.onPause(() => {
                    this.playicon = 'icon-bofang1';    });
                this.innerAudioContext.onEnded(() => {
                    this.playMusic(this.songId);       });
            }
        });
    }
```

➢ 在 onLoad 函数中调用 playMusic 方法,添加以下代码。

```
this.playMusic();
```

➢ 给碟片添加点击事件。实现点击碟片时,调用 handleToPlay 方法。

```
<view class="detail-play"    @tap="handleToPlay">
            <image :src="songDetail.al.picUrl" mode="" ></image>
            <text class="iconfont" :class="playicon"></text>
        </view>
```

➢ 在 methods 中添加 handToPlay 方法。

```
handleToPlay() {
    if (this.innerAudioContext.paused) {
        this.innerAudioContext.play();
    } else {
        this.innerAudioContext.pause();
```

➢ 实现碟片中图片的旋转功能。

在\<style>\</style>中添加样式。

```
@keyframes move {
    from {   transform:rotate(0deg);    }
    to {     transform:rotate(360deg);  }
}
.detail-play .detail-play-run {
    animation-play-state:running;
}
```

在图片上应用样式 detail-play-run。

```
<view class="detail-play"    @tap="handleToPlay">
        <image :src="songDetail.al.picUrl" mode="" class="detail-play-run" >
</image>
        <text class="iconfont" :class="playicon"></text>
</view>
```

此时碟片中的图片会旋转。

增加旋转控制变量 isRotate，该变量值为 Boolean 类型。暂停歌曲时，图片不旋转，isRotate 为 false；播放歌曲时，图片旋转，isRotate 为 true。将代码中的 "class="detail-play-run"" 改为 ":class="{'detail-play-run':isRotate}""，并修改\<script>\</script>模块的代码。

```
<script>
    export default {
        data() {
            return {
                ……省略
                playicon:'icon-pause',
                isRotate:true
            }
        },
        ……省略
        Methods{
            ……省略
            playMusic(songId) {
                ……省略
this.playicon = 'icon-bofang1';
                this.isRotate = false;
                uni.request({
                    ……省略
                    success:(res) => {
                        ……省略
                        this.innerAudioContext.onPlay(() => {
                            this.playicon = 'icon-pause';
                            isRotate:true;
                        });
                        this.innerAudioContext.onPause(() => {
                            this.playicon = 'icon-bofang1';
                            isRotate:false;
                        });
                        ……省略
```

```
            }
        });
    },
……省略
  }
}
</script>
```

（3）实现显示歌词功能

① JS 代码。

通过 list.vue 页面传递的歌曲"id"，请求地址 http://localhost:3000/lyric?id=歌曲 id 得到歌词，需将其格式进行转换。初始歌词格式如下。

"[00:00.000]作词 : 30 年前, 50 年后 [00:01.000]作曲 : 30 年前, 50 年后 [00:12.150]（每个身体, 情不自禁） [00:16.440]是现在我所有期待 [00:19.680]所有的爱 [00:21.720]为什么不明白 [00:25.200]说的话为什么不记得…"

转换后的数据为数组，格式如下。

[{lyric:" 作词 : 30 年前, 50 年后",time: "0.00"},
{lyric:" 作曲 : 30 年前, 50 年后",time: "1.00"},…]

在 data 部分添加变量 songLyric 和 lyricIndex，其中，songLyric 用来存储转换后的歌词，lyricIndex 用来存储当前播放歌曲的歌词。

```
songLyric:[],
lyricIndex:0,
```

在 methods 中添加方法 getLyricData 和 formatTimeToSec。formatTimeToSec 用于转换时间格式，getLyricData 用于获取原始歌词，并将其转换为数组存入 songLyric。

```
getLyricData() {
    uni.request({
    url:'http://localhost:3000/lyric?id=' + this.songid,
        method:'GET',
        success:(res) => {
            console.log("lyric success");
            console.log(res.data);
            let lyric = res.data.lrc.lyric;
            let result = [];
            let re = /\[(([^\]]+)\]([^[]+)/g;
            lyric.replace(re, ($0, $1, $2) => {
                result.push({
                time:this.formatTimeToSec($1),
                lyric:$2
                });
            });
            this.songLyric = result;
            console.log(result);
        },
        fail:() => {
            console.log("lyric 失败");
            }
        });
    },
```

```
formatTimeToSec(time) {
    var arr = time.split(':');
    return (parseFloat(arr[0]) * 60 + parseFloat(arr[1])).toFixed(2);
}
```

在 onLoad 函数中,添加 getLyricData 方法和 playMusic 方法的调用。

```
this.getLyricData();
this.playMusic();
```

在 methods 中添加监听歌词和取消监听歌词的方法。

```
listenLyricIndex() {
    clearInterval(this.timer);
    this.timer = setInterval(() => {
        for (var i = 0; i < this.songLyric.length; i++) {
            if (this.songLyric[this.songLyric.length - 1].time < this.innerAudioContext.currentTime) {
                this.lyricIndex = this.songLyric.length - 1;
                break;
            }
            if (this.songLyric[i].time < this.innerAudioContext.currentTime &&
                this.songLyric[i + 1].time > this.innerAudioContext.currentTime) {
                this.lyricIndex = i;
            }
        }
    });
},
cancelLyricIndex() {
    clearInterval(this.timer);
}
```

当歌曲处于播放状态时,需要监听歌词;当歌曲处于停止播放状态时,取消歌词监听。所以需要修改 playMusic 方法,添加监听和取消监听歌词的代码。

```
this.innerAudioContext.onPlay(() => {
    this.playicon = 'icon-pause';
    this.isRotate = true;
    this.listenLyricIndex();
});
this.innerAudioContext.onPause(() => {
    this.playicon = 'icon-bofang1';
    this.isRotate = false;
    this.cancelLyricIndex();
});
```

② 页面代码。

在碟片容器的下面添加歌词容器。

```
<view class="detail-lyric">
    <view class="detail-lyric-wrap"    :style="{ transform :'translateY(' + -(lyricIndex - 1) * 82 + 'rpx)' }">
        <view class="detail-lyric-item" :class="{ active :lyricIndex == index}" v-for="(item,index) in songLyric" :key="index">   {{ item.lyric }}   </view>
    </view>
</view>
```

在<style></style>模块添加 CSS 样式。

```css
.detail-lyric {
    height:246rpx;
    line-height:82rpx;
    font-size:32rpx;
    text-align:center;
    color:#959595;
    overflow:hidden;
}
.active {    color:white;    }
.detail-lyric-wrap {    transition:.5s;    }
.detail-lyric-item {        height:82rpx;    }
```

(4)实现显示评论功能

① <script></script>模块中的 JS 代码。

在 data 部分添加变量。

```js
data() {
    return {
    ……省略
        songComment:[]
        }
}
```

在 methods 中添加方法 MusicComment。这里将评论数组长度设置为 5，在页面中只显示 5 条评论。

```js
MusicComment() {
        uni.request({
            url:'http://localhost:3000/comment/music?id=' + this.songid ,
            method:'GET',
            success:(res) => {
                console.log("comment success");
                this.songComment = res.data.hotComments;
                this.songComment.length = 5;
                console.log(res.data.hotComments)
                }
            })
        }
```

在 onLoad 函数中，添加调用 MusicComment 的语句。

```js
onLoad(options) {
    this.songid = options.songid;
    this.playMusicDetail();
    this.innerAudioContext = null;
    this.getLyricData();
    this.playMusic();
    this.MusicComment();
}
```

② 页面代码

评论以列表的形式显示在歌词下面，如图 6-13 所示。

图 6-13 评论效果

在歌词容器下面添加下列代码。

```
    <view class="detail-comment">
      <view class="detail-comment-title">精彩评论</view>
      <view class="detail-comment-item" v-for="(item,index) in songComment" :key="index">
        <view class="detail-comment-img">
          <image :src="item.user.avatarUrl" mode=""></image>
        </view>
        <view class="detail-comment-content">
          <view class="detail-comment-head">
            <view class="detail-comment-name">
              <view>{{ item.user.nickname }}</view>
              <view>{{ item.time | formatTime }}</view>
            </view>
            <view class="detail-comment-like">
 {{ item.likedCount | formatNumber }} <text class="iconfont icon-like" style="color:pink;"></text>
            </view>
          </view>
          <view class="detail-comment-text">
            {{ item.content }}
          </view>
        </view>
      </view>
    </view>
```

在<style></style>模块中添加 CSS 代码。

```
.detail-comment {         margin:0 32rpx;      }
    .detail-comment-title {
      font-size:36rpx;
      color:white;
      margin:50rpx 0;
```

```css
}
.detail-comment-item {
    display:flex;
    margin-bottom:28rpx;
}
.detail-comment-img {
    width:66rpx;
    height:66rpx;
    border-radius:50%;
    overflow:hidden;
    margin-right:18rpx;
}
.detail-comment-img image {
    width:100%;
    height:100%
}
.detail-comment-content {
    flex:1;
    color:#cac9cd;
}
.detail-comment-head {
    display:flex;
    justify-content:space-between;
}
.detail-comment-name view:nth-child(1) {
    font-size:24rpx;
}
.detail-comment-name view:nth-child(2) {
    font-size:20rpx;
}
.detail-comment-like {
    font-size:30rpx;
}
.detail-comment-text {
    line-height:40rpx;
    color:white;
    font-size:28rpx;
    margin-top:16rpx;
    border-bottom:1px #595860 solid;
    padding-bottom:40rpx;
}
```

本案例主要实现了歌曲的播放功能，读者还可以自行实现歌曲搜索、视频播放等功能。

本章小结

本章主要介绍了 uni-app 的媒体控制、文件操作、设备操作、登录等 API，通过实例展示了多个 API 的使用方法。最后通过一个仿网易云音乐 App 的音乐播放器案例综合应用相关知识，在提升读者综合实践能力的同时，提升读者的音乐素养。

项目实战

1. 完善仿网易云音乐 App 的音乐播放器，实现视频展示页和视频播放页，如图 6-14 所示。除此之外，还可以实现搜索页。

2. 完善第 5 章的智云翻译项目，添加拍照翻译、语音翻译等功能。拍照翻译页如图 6-15 所示。

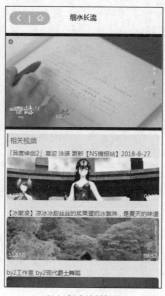

（a）视频展示页　　　（b）视频播放页

图 6-14　页面效果　　　　　　　　　　图 6-15　拍照翻译页

拓展实训项目

党的二十大报告指出"发展乡村特色产业，拓宽农民增收致富渠道""巩固拓展脱贫攻坚成果"。我国是世界上人口最多的发展中国家，农村经济发展不平衡现象突出。

贫困地区家副产品网络销售平台（又称"扶贫 832 平台"）是中国供销电子商务有限公司在财政部、国家乡村振兴局、中华全国供销合作总社的指导下，建设和运营的集"交易、服务、监管"功能于一体的平台。参考该平台，实现一个农副产品销售平台。

第7章
智慧环保项目

本章导读

本章以智慧环保项目为例介绍 uni-app 开发的流程，综合应用 uni-app 的组件、API、扩展组件 uView 等。智慧环保项目包括首页、回收分类页、分类查询结果页等各种页面。

学习目标

知识目标	1. 掌握 uView 2.0 的使用 2. 掌握 API 的使用 3. 掌握 flex 布局
能力目标	1. 能够查看 uView 的官方文档 2. 能够理解 scss 代码 3. 能够实现常规的移动页面开发 4. 具有使用 uView 开发实用项目的能力
素质目标	1. 具有积极探索新知、自主学习与钻研的精神 2. 具有节能控制、绿色发展的理念 3. 具有精益求精、自主解决问题的素质 4. 具有团队协作、善于沟通的素质

知识思维导图

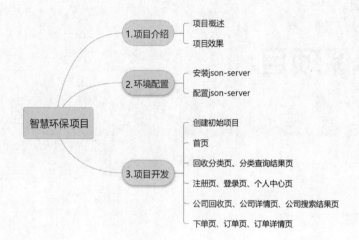

7.1 项目介绍

7.1.1 项目概述

如今，城市化速度越来越快，社会发展形态也在发生变化，因此，我国提出了创造智慧城市、提高城市管理水平以及提供多元化城市服务的发展战略。在此背景下，人们利用各种先进信息技术，有效整合城市各项配套系统和功能模块，进一步促进城市朝工业化、信息化、城镇化方向发展。党的二十大报告指出"加快构建废弃物循环利用体系""打造宜居、韧性、智慧城市"。城市环境保护工作有序开展，本项目实现一个简洁版的智慧环保系统，主要包括的页面如下。

智慧环保项目

（1）首页：展示环境宣传信息和快捷功能。

（2）订单页：登录后，用户可以查看订单信息。普通用户可以查询待接单、已接单、已完成、已取消的订单信息，并可以修改订单状态。企业用户可以查看相应的订单信息、处理客户提交的订单信息。

（3）公司回收页：用户可以查看附近的回收公司，还可以查看回收公司的详情。

（4）个人中心页：展示登录用户的个人信息。

（5）回收分类页：让用户了解哪些垃圾属于可回收的，以便于选择回收类型。

（6）分类查询结果页：显示回收分类页中点击"搜索"按钮后的结果。

（7）注册页和登录页：当处于需要登录才能查看的页面时，如未登录，会跳转到登录页。注册页用于注册新用户。

（8）下单页：选择需要回收的垃圾的类别，填写预约上门信息，让工作人员上门服务。

（9）公司搜索结果页：显示在公司回收页中点击"搜索"按钮后的结果。

（10）公司详情页：显示单个回收公司的详细信息。

（11）订单详情页：显示单个订单的详细信息。

7.1.2 项目效果

（1）首页

首页包括搜索框、轮播图、快捷功能、回收操作流程、【预约上门回收】按钮、爱心活动等，如图 7-1 所示。点击快捷功能图标，将进入对应页面。点击【预约上门回收】按钮，将进入公司回收页。

（2）订单页

在已登录的情况下，用户可在订单页查看所有待接单信息，如图 7-2 所示。点击某一订单，将进入相应的订单详情页。点击顶部的【待接单】【已接单】【已完成】【已取消】，将展示对应的订单列表。如果未登录，打开订单页会跳转到登录页。

图 7-1 首页

图 7-2 订单页

（3）公司回收页

该页面显示所有的回收公司信息，如图 7-3 所示。在此页面上方的搜索框输入相应关键字进入公司搜索结果页，点击某一公司信息，将进入相应的公司详情页。

（4）个人中心页

在未登录的情况下，个人中心页显示未登录。如已登录将显示用户的相关信息，如图 7-4 所示。点击【签到】，则可打开日历进行签到。点击快捷功能，将进入对应的功能页面。

图 7-3 公司回收页

图 7-4 个人中心页

（5）回收分类页

点击首页中的【回收分类】，进入回收分类页，如图 7-5 所示。在该页面左侧点击大类，该页面右侧显示相应的子类信息。

（6）分类查询结果页

在首页的搜索框或本页的搜索框中输入查询关键字进行查询，可进入分类查询结果页，如图 7-6 所示。

图 7-5 回收分类页

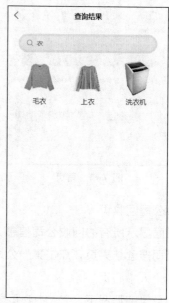

图 7-6 分类查询结果页

（7）注册页和登录页

要查看订单、预约上门回收等都必须先登录才可以进行。如果没有登录，则会跳转到登录页。如果没有注册，则会跳转到注册页。注册页如图 7-7 所示，登录页如图 7-8 所示。

图 7-7　注册页

图 7-8　登录页

（8）下单页

在公司回收页点击【立即下单】按钮，或在公司详情页中点击【预约上门回收】按钮，会跳转到下单页，如图 7-9 所示。在输入相关信息后，点击【立即预约】按钮，如果预约成功，则跳转到订单页，可查看提交的订单信息。

（9）公司搜索结果页

在公司回收页的搜索框中输入关键字进行搜索，可进入公司搜索结果页，如图 7-10 所示。

图 7-9　下单页

图 7-10　公司搜索结果页

（10）公司详情页

在公司回收页的公司列表中点击公司列表项，则会打开公司详情页，如图 7-11 所示。

（11）订单详情页

在订单页中点击订单列表项，则会打开订单详情页，如图 7-12 所示。

图 7-11 公司详情页

图 7-12 订单详情页

7.2 环境配置

7.2.1 安装 json-server

开发项目的时候，如果后端接口还没有创建出来，前端工程师也不必等后端接口创建出来才进行下一步开发。可以使用 mock.js 来模拟接口数据。这种用于模拟后端接口数据的工具有很多，例如 fastmock、lazy-mock、api-fox、json-srever 等。本项目采用 json-server 工具。json-server 是一个存储 JSON 数据的服务器，用来模拟后端接口数据。

在命令提示符窗口中，运行下列命令（要先安装 Node.js，参考 2.2 节）。

```
npm install -g json-server
```

可以通过查看版本号来测试 json-server 是否安装成功。

```
json-server -version
```

在任意地方新建一个文件夹（提供的源代码中为 smartEpserver 文件夹），进入该文件夹，在地址栏中输入"cmd"并按"Enter"键，则在命令提示符窗口中进入该文件夹。执行下列命令启动一个 Web 服务器，如图 7-13 所示。

```
json-server --watch db.json
```

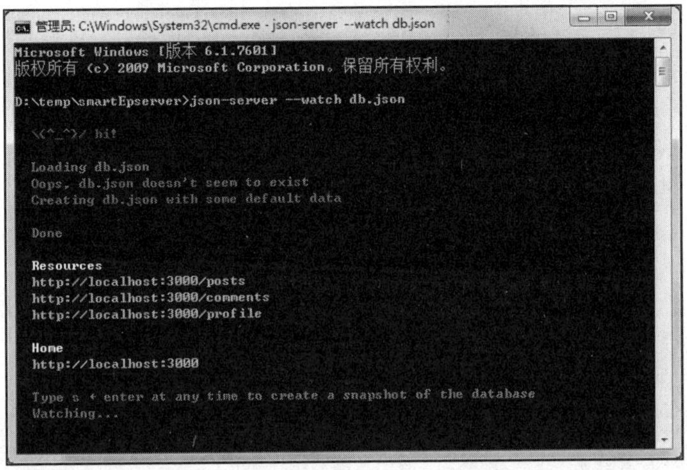

第 7 章 智慧环保项目

图 7-13 启动一个 Web 服务器

执行以上命令后,可以看到新建的文件夹中多了一个 db.json 文件,里面有一些默认数据。在浏览器中输入 http://localhost:3000,可以打开对应页面,如图 7-14 所示。这样服务器就建成了,里面有 3 个子页面地址,这就是模拟的接口地址。访问这些地址,可以显示 db.json 文件中对应的内容。

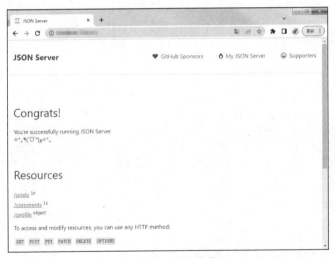

图 7-14 页面效果

7.2.2 配置 json-server

如果不想用 3000 端口,可以在启动服务器时指定需要的端口。以下命令用于将服务器端口改为 3004。

```
json-server --watch db.json --port 3004
```

在 smartEpserver 的根目录下新建一个 package.json 文件,将启动命令写到该文件中,代码如下。

```
{
    "scripts":{            "mock":"json-server db.json --port 3004"      }
}
```

以后启动服务器可用 npm run mock 命令。如果要停止服务器,只需关闭命令提示符窗口。

在 smartEpserver 的根目录下新建一个 public 文件夹,将图片资源放入 public 文件夹。修改 db.json 文件,具体代码如下所示。这里由于篇幅限制,每一种类型的记录仅保留了一条,其他省略。

```
{
  "types":[
  {    "id":1,
    "parentId":0,
    "icon":"",
    "name":"废纸",
    "unitPrice":0
  },
    ……省略

  {   "id":101,
    "parentId":1,
    "icon":"/image/type/101.jpg",
    "name":"报纸",
    "unitPrice":10
  },
    ……省略
  ],
  "users":[
{   "id":1,
    "phoneNum":"1567×××××××",
    "nickName":"xiaoyu",
    "password":"123456",
    "avatar":"/image/avatar/3.jpeg",
    "typeid":1
  },
    ……省略
  ],
  "hotDots":[
{    "id":1,
    "image":"/image/swt1.jpg",
    "title":""
  },
    ……省略
  ],
  "actions":[
{   "id":1,
    "content":"红牌洗衣液可以洗羊毛衫、羽绒服、棉织品等。",
    "coverUrl":"/image/xiyiye.jpg",
    "title":"200 积分限时兑换 500mL 浓缩洗衣液,快来抢购!"
  },
    ……省略
```

```
    ],
    "companys":[
      {  "id":1,
         "address":"湖北省武汉市光谷民族大道下线村100米处",
         "contact":"1841XXXXXXX",
         "coverUrl":"/image/company1.jpg",
         "introduction":"再生资源回收以物资不断循环……",
         "name":"信昌废旧物资回收公司",
         "score":1
      },
      ……省略
    ],
    "orders":[
      {  "id":1,
         "companyName":"时代回收公司",
         "createby":"小明",
         "createTime":"2022-03-15 10:35:58",
         "pickupTime":"14:00—16:00",
         "pickupAddress":"武汉市江夏区蜜糖镇",
         "goodNumber":3,
         "goodUnitPrice":2,
         "goodTypeName":"牛仔裤",
         "coverUrl":"/image/1.jpg",
         "contact":"1567××××××",
         "remark":"",
         "pkgs":12,
         "state":1,
         "companyId":3,
         "userId":2
      },
      ……省略
    ]
}
```

通过访问 http://localhost:3004/users 可以得到 users 里面的所有记录。若访问成功，返回的 res 的结构如图 7-15 所示，其中返回的数据 res.data 为数组。有关 json-server 更多的知识，请参考本书的电子资源。

▼{data: Array(1), statusCode: 200, header: {…}, errMsg: "request:ok"}

图 7-15　res 的结构

本项目的接口如表 7-1 所示。

表 7-1　接口

序号	名称	接口	请求方式
1	用户列表	http://localhost:3004/users	GET
2	订单列表	http://localhost:3004/orders	GET
3	分类列表	http://localhost:3004/types	GET

续表

序号	名称	接口	请求方式
4	轮播宣传图片列表	http://localhost:3004/hotDots	GET
5	活动列表	http://localhost:3004/actions	GET
6	公司列表	http://localhost:3004/companys	GET
7	注册用户	http://localhost:3004/users	POST
8	下单	http://localhost:3004/orders	POST
9	取消订单	http://localhost:3004/orders/订单编号	PATCH

7.3 项目开发

7.3.1 创建初始项目

（1）新建项目

用默认模板新建一个 Vue2 的 uni-app 项目 smartEpApp。新建 3 个 tabBar 页面（已自动生成一个 index.vue 页面），分别为订单页 order.vue、公司回收页 company.vue、个人中心页 me.vue。修改 pages.json 文件，增加 tabBar 节点，实现底部导航栏。pages.json 的代码如下。

```
{
    "pages":[{
        "path":"pages/index/index",
        "style":{    "navigationBarTitleText":"诚 信 回 收",
            "navigationBarBackgroundColor":"#00c297",
            "navigationBarTextStyle":"white"
        }
    }, {
        "path":"pages/company/company",
        "style":{    "app-plus":{
            "titleNView":false
            }
        }
    }, {
        "path":"pages/order/order",
        "style":{    "navigationBarTitleText":"我的订单",
            "enablePullDownRefresh":false
        }

    }, {
        "path":"pages/me/me",
        "style":{    "navigationBarTitleText":"个人中心"
        }

    }],
    "globalStyle":{
"navigationBarTextStyle":"black",
        "navigationBarTitleText":"uni-app",
```

```
        "navigationBarBackgroundColor":"#F8F8F8",
        "backgroundColor":"#F8F8F8",
        "app-plus":{    "background":"#efeff4"     }
    },
    "tabBar":{
        "color":"#909399",
        "selectedColor":"#303133",
        "backgroundColor":"#FFFFFF",
        "borderStyle":"black",
        "list":[{
            "pagePath":"pages/index/index",
            "iconPath":"static/tabbar/home.png",
            "selectedIconPath":"static/tabbar/home-selected.png",
            "text":"首页"
        },{
            "pagePath":"pages/order/order",
            "iconPath":"static/tabbar/order.png",
            "selectedIconPath":"static/tabbar/order-selected.png",
            "text":"订单"
        },{
            "pagePath":"pages/company/company",
            "iconPath":"static/tabbar/company.png",
            "selectedIconPath":"static/tabbar/company_selected.png",
            "text":"公司回收"
        },   {
            "pagePath":"pages/me/me",
            "iconPath":"static/tabbar/user.png",
            "selectedIconPath":"static/tabbar/user-selected.png",
            "text":"我的"
        } ]
    }
}
```

其中，company.vue 页面使用代码""app-plus": {"titleNView": false}"隐藏了导航栏。可以通过设置""navigationStyle": "custom""，使用自定义导航栏，以关闭默认的原生导航栏。

```
{
    "path":"pages/company/company",
    "style":{
        "navigationStyle":"custom"
    }
}
```

下面将其他页面配置一并放到 pages 中。

```
"pages":[
.      ……省略前面的 tabBar 页面
,{    "path":"childPages/type/recyclingMenu",
      "style":{    "navigationBarTitleText":"",
            "enablePullDownRefresh":false
      }
},{   "path":"childPages/type/searchRes",
      "style":{    "navigationBarTitleText":"",
            "enablePullDownRefresh":false
```

```json
    },{ "path":"pages/login/login",
        "style":{    "navigationBarTitleText":"",
            "enablePullDownRefresh":false
        }
    },{ "path":"pages/login/register",
        "style":{    "navigationBarTitleText":"",
            "enablePullDownRefresh":false
        }
    },{ "path":"childPages/company/detail",
        "style":{    "navigationBarTitleText":"诚信回收",
            "navigationBarBackgroundColor":"#00c297",
            "navigationBarTextStyle":"white"
        }
    },{ "path":"childPages/company/searchRes",
        "style":{    "navigationBarTitleText":"搜索结果",
            "navigationBarBackgroundColor":"#00c297",
            "navigationBarTextStyle":"white"
        }
    },{ "path":"childPages/company/fillIn",
        "style":{    "navigationBarTitleText":"",
            "enablePullDownRefresh":false
        }
    },,{   "path" :"childPages/order/waiting",
        "style" :{    "navigationBarTitleText":"",
            "enablePullDownRefresh":false
        }
    }]
```

（2）安装插件

在插件市场搜索前端组件 uView，使用 HBuilderX 导入插件的方式安装 uView 2.0。安装完成后，项目根目录下多了一个 uni_modules 文件夹，uView 的组件都在/uni_modules/uview-ui 文件下面。

uView 依赖 scss，必须安装 scss 插件，uView 才能正确运行。在插件市场找到 "scss/sass 编译"插件，进行安装。如果已安装，则略过此步。

（3）配置 uView

引入 uView 主 JS 库。在项目根目录中的 main.js 中，添加以下代码，引入并使用 uView 的主 JS 库。需要注意的是，这两行代码要放在"import Vue from 'vue'"的后面。

```
import uView from '@/uni_modules/uview-ui'
Vue.use(uView)
```

引入主题文件。在项目根目录下的 uni.scss 文件中引用 uView 的全局 scss 主题文件 theme.scss。

```
@import '@/uni_modules/uview-ui/theme.scss';
```

引入 uView 样式文件。在 App.vue 中的<style></style>部分，引入 uView 基础样式文件 index.scss。注意，如果还引用了其他样式文件，引用该样式文件的代码需要写在第 1 行，同时在<style></style>标签中添加" lang="scss""。具体代码如下。

```
<style lang="scss">
    @import "@/uni_modules/uview-ui/index.scss";/* 注意要写在第1行 */
</style>
```

配置 easycom。在项目根目录的 pages.json 中添加 easycom 字段。如果是通过 HBuiderX 导入插件的方式安装的 uView，可以忽略此配置。

```
"easycom":{
    "^u-(.*)":"@/uni_modules/uview-ui/components/u-$1/u-$1.vue"
},
```

（4）配置全局 baseUrl

在 smartEpApp 根目录下新建 common 文件夹，在 common 文件夹下新建文件 api.js，代码如下。

```
const baseUrl = 'http://localhost:3004';
export default {    baseUrl }
```

将服务器 URL 在 main.js 中进行全局挂载。在 main.js 中创建 Vue 实例之前，添加以下代码。

```
import baseUrl from '@/common/api.js'
Vue.prototype.$baseUrl = baseUrl
```

这样在页面中可以用 this.$baseUrl.baseUrl 来获取 http://localhost:3004。

（5）页面的文件路径如表 7-2 所示。

表 7-2 页面的文件路径

序号	页面	文件路径
1	首页	pages/index/index.vue
2	订单页	pages/order/order.vue
3	公司回收页	pages/company/company.vue
4	个人中心页	pages/me/me.vue
5	回收分类页	childPages/type/recyclingMenu.vue
6	分类查询结果页	childPages/type/searchRes.vue
7	登录页	pages/login/login.vue
8	注册页	pages/login/register.vue
9	公司详情页	childPages/company/detail.vue
10	公司搜索结果页	childPages/company/searchRes.vue
11	下单页	childPages/company/fillIn.vue
12	订单详情页	childPages/order/waiting.vue

7.3.2 首页

首页共有 6 个部分：搜索框、轮播图、快捷功能、回收操作流程、【预约上门回收】按钮、爱心活动。轮播图、爱心活动的数据来源于后端服务器，效果如图 7-1 所示，其请求地址如下，具体数据格式参见 7.2.2 小节的 db.json 文件。

> 轮播图的请求地址：http://localhost:3004/hotDots。
> 爱心活动信息的请求地址：http://localhost:3004/actions。

首页完整代码如下。

```
<template>
    <view>
        <view class="search">
            <u-search placeholder="请输入搜索内容" height="50" shape="square" v-model="keyword" @search="search"   :showAction="false"></u-search>
        </view>
        <view class="swiper">
            <u-swiper :list="swiperlist" keyName="image" :autoplay="true" circular></u-swiper>
        </view>
        <view class="char">
            <view class="char_item" @click="goto(1)">
                <image mode="widthFix" src="../../static/icons/Recycling_classification.png"></image>
                <view>回收分类</view>
            </view>
            <view class="char_item">
                <image mode="widthFix" src="../../static/icons/Old_things.png"></image>
                <view>旧物去向</view>
            </view>
            <view class="char_item">
                <image mode="widthFix" src="../../static/icons/Recycling_machine.png"></image>
                <view>附近回收</view>
            </view>
            <view class="char_item">
                <image mode="widthFix" src="../../static/icons/Shopping_Mall.png"></image>
                <view>积分商城</view>
            </view>
        </view>
        <view class="btnOrder">
            <u-button color="rgb(0, 194, 151)" size="large" shape="circle" @click="goto(5)">预约上门回收</u-button>
        </view>
        <view class="char char_m">
            <view class="char_item">
                <u-icon name="clock" size="20" label="在线预约"></u-icon>
                <view>第1步</view>
            </view>
            <view class="char_item">
                <u-icon name="bell" size="20" label="免费上门"></u-icon>
                <view>第2步</view>
            </view>
            <view class="char_item">
                <u-icon name="attach" size="20" label="订单完成"></u-icon>
                <view>第3步</view>
```

```html
            </view>
            <view class="char_item">
                <u-icon name="star" size="20" label="用户福利"></u-icon>
                <view>第 4 步</view>
            </view>
        </view>
        <view class="activeWrap">
            <view class="title">爱心活动</view>
            <view class="actlist">
                <view class="act_item" v-for="(item,index) in activelist" :key="index">
   <u-image :src="$baseUrl.baseUrl+item.coverUrl" mode="aspectFill" :border-radius="20" width="100%"    height="300rpx"></u-image>
                    <text>{{item.title}}</text>
                </view>
            </view>
        </view>
    </view>
</template>
<script>
    export default {
        data() {
            return {
                swiperlist:[],
                keyword:'',
                activelist:[]
            }
        },
        onLoad() {
            this.getHotDots();
            this.getActlist();
        },
        methods:{
            getHotDots() {
                uni.request({
                    url:this.$baseUrl.baseUrl + '/hotDots',
                    methods:'get',
                    success:(res) => {
                        console.log(res.data);
                        this.swiperlist = res.data;
                        for (let i = 0; i < this.swiperlist.length; i++) {
                    this.swiperlist[i].image = this.$baseUrl.baseUrl + this.swiperlist[i].image;
                        }
                    }
                })
            },
            search(e) {
                console.log(e);
                uni.navigateTo({
                    url:'/childPages/type/searchRes?txtSearch=' + e
                });
                this.keyword = "";
```

```
            },
            goto(i) {
                switch (i) {
                    case 1:
                        uni.navigateTo({    url:'/childPages/type/recyclingMenu'    });
                        break;
                    case 5:
                        uni.switchTab({     url:'/pages/company/company'    })
                        break;
                    default:
                        break;
                } },
            getActlist() {
                uni.request({
                    url:this.$baseUrl.baseUrl + '/actions',
                    methods:"get",
                    success:(res) => {
                        console.log(res);
                        this.activelist = res.data;
                        console.log(this.activelist);
                    }
                })
            }
        }
    }
</script>
<style lang="scss" scoped>
    .search {
        width:100%;
        box-sizing:border-box;
        padding:20px 20px 15px 20px;
        background-color:rgb(0, 194, 151);
    }
    .swiper {
        padding:10px;
        background-color:#F4F4F5;
    }
    .char {
        margin-top:20px;
        margin-left:10rpx;
        display:flex;
        justify-content:space-around;
    }
    .char_item {
        flex:1;
        text-align:center;
        font-size:14px;
        color:#888888;
    }
    .char_item image {
        width:60%;
    }
    .btnOrder   {padding:20px;     }
    .char_m {    margin-top:0;     }
```

```
        .activeWrap {     padding:0 20px;      }
        .activeWrap .title {
            font-size:16px;
            padding-left:10px;
            border-left:3px solid rgb(0, 194, 151);
            margin-top:10px;
            margin-bottom:10px;
        }
        .activeWrap .actlist {    width:100%;    }
        .act_item {        margin-bottom:10px;  }
</style>
```

【代码解析】

（1）搜索框

搜索框使用了 uView 中的 search 组件，该组件双向绑定了 data 属性 keyword，按"Enter"键或者手机软键盘右下角的"搜索"键时，触发@search 事件。search 组件可以开启右边按钮，可将该按钮设置为"搜索"或"取消"等内容。:showAction="false"表示右侧的"搜索"按钮不显示。

```
<u-search placeholder="请输入搜索内容" height="50" shape="square" v-model=
"keyword" @search="search"    :showAction="false"></u-search>
```

JS 代码中，在 data 处添加属性 keyword:''，在 methods 处添加 search 方法。

（2）轮播图

轮播图使用了 uView 中的 u-swiper 组件。其中 list 属性值为轮播数据，keyName 为 list 数组中对象的目标属性名，autoplay 用于设置是否自动播放，circular 用于设置是否循环播放。

```
<u-swiper :list="swiperlist" keyName="image" :autoplay="true" circular>
</u-swiper>
```

JS 代码中，在 data 处添加属性 swiperlist:[ ]，在 methods 处添加 getHotDots，onLoad 生命周期函数调用 getHotDots 方法。

（3）快捷功能

这里有 4 个快捷，若点击【回收分类】，执行 goto 方法，跳转到回收分类页。因为后面的代码还没有实现，读者在调试时，应先将对应代码注释掉。快捷功能在布局上使用了 flex 布局。

（4）【预约上门回收】按钮

该按钮使用了 u-Button 组件，color 属性用于设置背景颜色；type 属性用于设置不同颜色的按钮样式，如 default、primary、success、info、waring、error；size 用于设置按钮大小，这里将其设置为 large（大按钮）；shape 用于设置按钮是 circle（圆角）还是 square（直角）。点击此按钮执行 goto 方法。

（5）回收操作流程

与快捷功能一样，回收操作流程使用 flex 布局。这里用了 u-icon 图标，u-icon 是基于字体的图标集，包含大多数场景的图标。其中，name 表示图标名称，读者可以在 uni-app 官网查看图标集。size 用于设置图标大小，默认单位为 px，默认值为 16px。label 用于设置图标周围的文字，可以用 labelPos 设置文字的位置，其取值为 left、top、right、bottom。

```
<u-icon name="clock" size="20" label="在线预约"></u-icon>
```

（6）爱心活动

爱心活动的数据来源于后端服务器，具体的数据格式可查看 7.2.2 小节的 db.json 文件。爱心活动的效果如图 7-16 所示。这里显示图片使用了 uView 的 u-image 组件，其中图片的 URL 是一个相对地址，其前面需要添加"http://localhost:3004"，可用全局属性$baseUrl.baseUrl 获得。

```
<u-image :src="$baseUrl.baseUrl+item.coverUrl" mode="aspectFill" :border-radius="20" width="100%"    height="300rpx"></u-image>
```

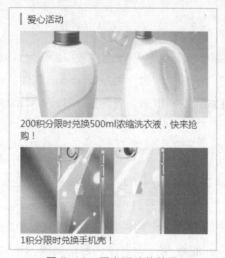

图 7-16　爱心活动的效果

7.3.3　回收分类页、分类查询结果页

这两个页面都用于对分类进行操作，是首页的子页面。这里将页面放在 childPages/type 目录下，为项目的分包处理提供支持，参见 2.1.3 小节。

回收分类页的效果如图 7-5 所示，分类查询结果页的效果如图 7-6 所示。分类的数据来源于后端服务器，数据格式请参考 7.2.2 小节中 db.json 文件的 types 属性。其 URL 为 http://localhost:3004/types。

1. 回收分类页

对于回收分类页的手风琴式的导航效果，uView 中提供了一个模板。读者可以在 HBuilderX 插件安装页面搜索"uView"，通过导入示例来获得相关代码，然后在此基础上根据数据格式进行修改，以显示回收分类。除此之外还在页面顶部添加了搜索框。它与首页中的搜索框类似，只是为了页面的美观，在其外面添加 view 组件，设置 CSS 样式。该页面的完整代码如下。

```
<template>
    <view class="u-wrap">
        <view class="u-search-box">
            <view class="u-search-inner">
```

```
                    <u-search placeholder="请输入搜索内容" height="30" shape="round" v-model="keyword" @search="search"    :showAction="false"></u-search>
                </view>
            </view>
            <view class="u-menu-wrap">
                <scroll-view scroll-y scroll-with-animation class="u-tab-view menu-scroll-view" :scroll-top="scrollTop">
                    <view v-for="(item,index) in parentlist" :key="index" class="u-tab-item" :class="[current==index ? 'u-tab-item-active' :'']"
                      :data-current="index" @tap.stop="swichMenu(index)">
                        <text class="u-line-1">{{item.name}}</text>
                    </view>
                </scroll-view>
                <block v-for="(item,index) in parentlist" :key="index">
                    <scroll-view scroll-y class="right-box" v-if="current==index">
                        <view class="page-view">
                            <view class="class-item">
                                <view class="item-title">
                                    <text>{{item.name}}</text>
                                </view>
                                <view class="item-container">
                                    <view class="thumb-box" v-for="(item1, index1) in childlist[current]" :key="index1">
                                        <image class="item-menu-image" :src="$baseUrl.baseUrl+item1.icon" mode=""></image>
                                        <view class="item-menu-name">{{item1.name}}</view>
                                    </view>
                                </view>
                            </view>
                        </view>
                    </scroll-view>
                </block>
            </view>
        </view>
</template>
<script>
    export default {
        data() {
            return {
                scrollTop:0, //左侧菜单的滚动条位置
                current:0, // 预设当前菜单项的值
                menuHeight:0, // 左侧菜单的高度
                menuItemHeight:0, // 左侧菜单项的高度
                parentlist:[], //左侧菜单的数据
                childlist:[],//右侧的数据与左侧菜单的数据通过数组索引进行对应
                keyword:''
            }
        },
        created() {
            this.getTypelist();
        },
        methods:{
            search(e) {
```

```js
        console.log(e);
        uni.navigateTo({
            url:'/childPages/type/searchRes?txtSearch=' + e
        });
        this.keyword = "";
    },
    getTypelist() {
        console.log(11);
        uni.request({
            url:this.$baseUrl.baseUrl + "/types?parentId=0",
            method:"GET",
            success:(res)=>{
                if (res.statusCode == 200) {
                    this.parentlist = res.data;
                    for(let i=0; i<this.parentlist.length; i++){
                        this.getChildlist(this.parentlist[i].id);
                    }
                }
            }
        })
    },
    getChildlist(id) {
        uni.request({
            url:"http://localhost:3004/types?parentId="+id,
            method:"GET",
            success:(res)=>{
                console.log(res.data);
                if (res.statusCode == 200) {
                    this.childlist.push(res.data);
                }
            }
        })
    },
    getImg() {
        return Math.floor(Math.random() * 35);
    },
    // 点击左边的菜单项切换回收分类
    async swichMenu(index) {
        if(index == this.current) return ;
        this.current = index;
        // 如果菜单或菜单项的高度为0，意味着尚未初始化
        if(this.menuHeight == 0 || this.menuItemHeight == 0) {
            await this.getElRect('menu-scroll-view', 'menuHeight');
            await this.getElRect('u-tab-item', 'menuItemHeight');
        }
        // 将菜单中的菜单项垂直居中
        this.scrollTop = index * this.menuItemHeight + this.menuItemHeight / 2 - this.menuHeight / 2;
    },
    // 获取一个目标元素的高度
    getElRect(elClass, dataVal) {
        new Promise((resolve, reject) => {
            const query = uni.createSelectorQuery().in(this);
```

```
                    query.select('.' + elClass).fields({size:true},res => {
                        // 如果节点尚未生成, res 值为 null, 循环执行
                        if(!res) {
                            setTimeout(() => {
                                this.getElRect(elClass);
                            }, 10);
                            return ;
                        }
                        this[dataVal] = res.height;
                    }).exec();
                })
            }
        }
    }
</script>
<style lang="scss" scoped>
    .u-wrap {
        height:calc(100vh);
        /* #ifdef H5 */
        height:calc(100vh - var(--window-top));
        /* #endif */
        display:flex;
        flex-direction:column;
    }
    .u-search-box { adding:18rpx 30rpx;    }
    .u-search-inner {
        background-color:rgb(234, 234, 234);
        border-radius:100rpx;
        display:flex;
        align-items:center;
        padding:10rpx 10rpx;
    }
    .u-menu-wrap {
        flex:1;
        display:flex;
        overflow:hidden;
    }
    .u-tab-view {
        width:200rpx;
        height:100%;
    }
    .u-tab-item {
        height:110rpx;
        background:#f6f6f6;
        box-sizing:border-box;
        display:flex;
        align-items:center;
        justify-content:center;
        font-size:26rpx;
        color:#444;
        font-weight:400;
        line-height:1;
    }
    .u-tab-item-active {
```

```css
    position:relative;
    color:#000;
    font-size:30rpx;
    font-weight:600;
    background:#fff;
}
.u-tab-item-active::before {
    content:"";
    position:absolute;
    border-left:4px solid rgb(0, 194, 151);
    height:32rpx;
    left:0;
    top:39rpx;
}
.u-tab-view {    height:100%;    }
.right-box {    background-color:rgb(250, 250, 250);    }
.page-view {    padding:16rpx;    }
.class-item {
    margin-bottom:30rpx;
    background-color:#fff;
    padding:16rpx;
    border-radius:8rpx;
}
.item-title {
    font-size:26rpx;
    color:$uni-text-color;
    font-weight:bold;
}
.item-menu-name {
    font-weight:normal;
    font-size:24rpx;
    color:$uni-text-color;
}
.item-container {
    display:flex;
    flex-wrap:wrap;
}
.thumb-box {
    width:33.333333%;
    display:flex;
    align-items:center;
    justify-content:center;
    flex-direction:column;
    margin-top:20rpx;
}
.item-menu-image {
    width:120rpx;
    height:120rpx;
}
</style>
```

【代码解析】

回收分类页的层级关系通过 parentId 来体现。其中 parentId 为 0 的项为左侧菜单项,

parentId 不为 0，则表示该项为 id 为 parentId 类的子项。这里首先查询 parentId 为 0 的项，将它们存储在 parentlist 数组中，同时查询该项的子项，将其所有的子项存储在 childlist 数组中，childlist 为一个二维数组。子项和其父项通过数组索引进行关联。

在 JS 代码中，getChildlist(id)方法用于将对应子项存入 childlist 数组。getTypelist 方法用于查找 parentId 为 0 的项，然后调用 getChildlist 查找其子项。

2．分类查询结果页

分类查询结果页显示包含关键字的所有子项和包含关键字的大类的子项，如图 7-6 所示。当搜索不到相关子项时，显示"暂无搜索数据！"的提示，如图 7-17 所示。

图 7-17 搜索不到相关子项时的演示效果

分类查询结果页的完整代码如下。

```
<template>
    <view class="page-view">
        <view class="u-search-box">
            <view class="u-search-inner">
                <u-search placeholder="请输入搜索内容" height="30" shape="round" v-model="keyword" @search="search"       :showAction="false"></u-search>
            </view>
        </view>
        <view class="class-item">
            <view class="item-container" v-if="childlist.length>0">
                <view class="thumb-box" v-for="(item,index) in childlist"  :key="index" >
                    <image mode="widthFix" :src="$baseUrl.baseUrl+item.icon"></image>
                    <view class="item-menu-name">{{item.name}}</view>
                </view>
            </view>
            <view class="emptyBox" v-else>
                <u-empty  text="暂无搜索数据！" mode="data"></u-empty>
            </view>
        </view>
    </view>
</template>
<script>
    export default {
        data() {
            return {
                childlist:[],
                keyword:''
            }
        },
        onLoad(options) {
            console.log(options.txtSearch)
```

```javascript
            this.getTypelist(options.txtSearch);
            this.keyword = options.txtSearch;
        },
        methods:{
    search(e) {
        console.log(e);
        uni.navigateTo({
            url:'/childPages/type/searchRes?txtSearch=' + e
        });
        this.keyword = "";
    },
            getTypelist(str) {
                uni.request({
                    url:this.$baseUrl.baseUrl + "/types?name_like="+str,
                    method:"GET",
                    success:(res)=>{
                    console.log(res.data);
                    if(res.statusCode == 200 && res.data.length>0) {
                        let i = 0;
                        for(i = 0; i< res.data.length;i++){
                        if(res.data[i].parentId == 0) {
                            this.getChildlist(res.data[i].id);
                        } else {
                            this.childlist.push(res.data[i]) ;
                        }
                        }
                    }
                    }
                })
            },
            getChildlist(id) {
                uni.request({
                    url:"http://localhost:3004/types?parentId="+id,
                    method:"GET",
                    success:(res)=>{
                    if(res.statusCode == 200) {
                        for(let i = 0 ;i< res.data.length; i++){
                            this.childlist.push(res.data[i]);
                        }
                    }
                    }
                });
            }
        }
    }
</script>
<style scoped>
    .page-view{    padding:20rpx;    }
    .u-search-box {    padding:18rpx 30rpx;    }
    .u-search-inner {
        background-color:rgb(234, 234, 234);
        border-radius:100rpx;
```

```
        display:flex;
        align-items:center;
        padding:10rpx 10rpx;
    }
    .class-item{
        margin-bottom:30rpx;
        background-color:#fff;
        padding:16rpx;
        border-radius:8rpx;
    }
    .item-container{
        display:flex;
        flex-wrap:wrap;
    }
    .thumb-box{
        width:33.33%;
        text-align:center;
    }
    .thumb-box image{    width:70%;   }
    .emptyBox{
        position:fixed;
        top:50%;
        left:50%;
        transform:translate3d(-50%,-50%,0);
    }
</style>
```

【代码解析】

在<template></template>中若 childlist.length > 0，则显示搜索到的所有子项，否则显示"暂无搜索数据！"。

7.3.4 注册页、登录页、个人中心页

1. 注册页

注册页用来注册新用户，用户类型分普通用户和企业用户，页面效果如图 7-7 所示。本实例主要实现的是普通用户的功能。注册成功会弹出一个提示框，然后自动登录系统，并打开个人中心页。

注册功能用 post 方法请求 http://localhost:3004/users。

```
<template>
    <view class="registerWrap">
        <!-- 标题 -->
        <view class="title">
            <h3>注册</h3>
            <h4>欢迎使用智慧环保</h4>
        </view>
        <!-- 表单 -->
        <view class="formWrap">
            <view class="formItem">
```

```html
            <view class="label">手机号:</view>
            <view class="inputBox">
                <input type="number" v-model="phoneNum" />
            </view>
        </view>
        <view class="formItem">
            <view class="label">昵称:</view>
            <view class="inputBox">
                <input type="text" v-model="nickName" />
            </view>
        </view>
        <view class="formItem">
            <view class="label">密码:</view>
            <view class="inputBox">
                <input password="true" v-model="password" />
            </view>
        </view>
    </view>
    <view>
<u-button type="primary" @click="submitReg" color="rgb(0,194,151)">注 册</u-button>
    </view>
    <navigator url="/pages/me/login">
        <view class="registTips"  >已有账号，立即登录</view>
    </navigator>
    <u-toast ref='uToast'/>
</view>
</template>
<script>
    export default {
        data() {
            return{
                flag:true,
                phoneNum:'',
                nickName:'',
                password:''
            }
        },
        methods:{
            isSame(){
                
                uni.request({
                    url:"http://localhost:3004/users",
                    data:{
                        phoneNum:this.phoneNum
                    },
                    success:(res) => {
                        if(res.data.length != 0){
                            this.flag = false;
                            uni.showToast({
                                title:'手机号已注册!! ',
                                icon:'error'
```

```
                })
            }
        }
    });
    console.log(this.flag);
    return this.flag;
},
submitReg() {
    if (this.phoneNum.length == 0) { //判断是否输入手机号
        //提示
        uni.showToast({
            title:'请先输入手机号!',
            icon:'error'
        })
    } else if (this.phoneNum.length != 11) {
        uni.showToast({
            title:'请输入正确手机号!',
            icon:'error'
        })
    } else if (this.nickName.length == 0) { //判断是否输入昵称
        uni.showToast({
            title:'请输入昵称!',
            icon:'error'
        })
    } else if (this.password.length == 0) { //判断是否输入密码
        uni.showToast({
            title:'请输入密码!',
            icon:'error'
        })
    } else if(this.isSame()) {
        //请求
        let URL = this.$baseUrl.baseUrl+'/users';
        uni.request({
            url:URL,
            data:{
                phoneNum:this.phoneNum,
                nickName:this.nickName,
                password:this.password,
                avatar:"/image/avatar/" + Math.ceil(Math.random()*10) + '.jpeg',
                typeid:1
            },
            method:'POST',
            header:{
                'content-type':'application/x-www-form-urlencoded',
            },
            success:(res)=>{
                this.$refs.uToast.show({
                    title:'登录成功',
                    type:'success'
                }),
```

```
                    this.autoLogin()   ;
                },
                fail:function(res) {
                    console.log("失败",res)
                }
            });
        }
    },
    autoLogin(){
        uni.request({
            url:this.$baseUrl.baseUrl+'/users',
            method:"GET",
            data:{
                phoneNum:this.phoneNum,
                password:this.password,
                typeId:this.typeid
            },
            success:(res) => {
                console.log(res);
                if (res.statusCode == 200) {
                    uni.setStorageSync('info', res.data[0]);
                    uni.switchTab({
                        url:'/pages/me/me'
                    });
                }
            }
        });
    }
}
}
</script>
<style scoped>
    .registerWrap{
        width:100%;
        box-sizing:border-box;
        padding:0 40rpx;
    }
    .title { padding-top:60rpx;      }
    .title h3{
        font-size:24px;
        line-height:40px;
        color:#333;
    }
    .title h4{
        font-size:18px;
        color:#888;
    }
    .formWrap {    margin-top:20px; }
    .formItem{
        padding-bottom:15px;
    }
    .formItem .label{
        font-size:14px;
```

```
            color:#333;
        }
        .inputBox{
            margin-top:5px;
            border:1px #c3c3c3 solid;
            border-radius:8px;
        }
        .inputBox input{
            height:38px;
            line-height:38px;
            text-indent:1em;
        }
        .registTips{
            padding-top:20px;
            text-align:center;
        }
</style>
```

【代码解析】

（1）随机图像

在提交之前，使用 JS 代码进行验证，当验证通过后进行注册，注册的数据头像 avatar 的值为：'/image/avatar/' + Math.ceil(Math.random()*14) + '.jpeg'。在服务器的 public/image/avatar 目录中存放了 14 张 JPEG 图片。使用"Math.ceil(Math.random()*14)"随机得到 1～14 的整数。

（2）自动登录、注册方法

autoLogin 方法：实现自动登录功能，若登录成功，则调用"uni.setStorageSync('info', res.data[0]);"将用户信息存入缓存，并跳转到个人中心页。

submitReg 方法：实现注册功能，若注册成功，则调用 autoLogin 进行自动登录，并跳转到个人中心页。

（3）轻量弹框

注册页使用了两种轻量弹框，一种是 uni-upp 的内置组件 Toast，另一种是 uView 中的 u-Toast 组件。在页面的最后放置了一个轻量弹框"<u-toast ref='uToast' />"，在 JS 中通过"this.$refs.uToast"得到该组件实例。

2. 登录页

要使用下单功能、查看订单状态等都需要处于登录状态。登录成功，则将用户信息存入缓存。登录页效果如图 7-8 所示。页面完整的代码如下。

```
<template>
    <view class="loginWrap">
        <view class="logo">
            <image src="../../static/logo.png" mode="widthFix"></image>
        </view>
        <view class="title">
            <h3>登录</h3>
            <h4>欢迎使用智慧环保</h4>
        </view>
        <view class="formWrap">
            <view style="margin-bottom:10rpx;">
```

```html
                <u-radio-group v-model="roleVal" placement="row">
                    <u-radio label="普通用户" name="普通用户" :customStyle="{marginRight:'16px'}" activeColor="rgb(0,194,151)"></u-radio>
                    <u-radio label="企业用户" name="企业用户" activeColor="rgb(0,194,151)"></u-radio>
                </u-radio-group>
            </view>
            <view class="formItem">
                <view class="label">手机号:</view>
                <view class="inputBox">
                    <input type="number" v-model="phoneNum" />
                </view>
            </view>
            <view class="formItem">
                <view class="label">密码:</view>
                <view class="inputBox">
                    <input type="number" password v-model="password" />
                </view>
            </view>
        </view>
        <view>
            <u-button type="primary" @click="loginSubmit" color="rgb(0,194,151)" >登 录</u-button>
        </view>
        <navigator url="register1">
            <view class="registTips" @tap="gotoReg">注册新用户</view>
        </navigator>
        <u-toast ref='uToast' />
    </view>
</template>
<script>
    export default {
        data() {
            return {
                phoneNum:'',
                password:'',
                roleVal:'普通用户',
                coverUrl:''
            }
        },
        methods:{
            gotoReg() {
                uni.navigateTo({
                    url:"/pages/login/register"
                })
            },
            loginSubmit() {
                if (this.phoneNum.length == 0) { //判断是否输入手机号
                    //提示
                    uni.showToast({
                        title:'请先输入手机号!',
```

```
                    icon:'error'
                })
            } else if (this.phoneNum.length != 11) {
                uni.showToast({
                    title:'请输入正确手机号!',
                    icon:'error'
                })
            } else if (this.password.length == 0) { //判断是否输入密码
                uni.showToast({
                    title:'请输入密码!',
                    icon:'error'
                })
            } else {
                //请求
                uni.request({
                    url:this.$baseUrl.baseUrl+'/users',
                    method:"GET",
                    data:{
                        phoneNum:this.phoneNum,
                        password:this.password,
                        typeid:this.roleVal == '普通用户' ? 1 :2
                    },
                    success:(res) => {
                        if (res.data.length > 0) {
                            console.log("成功:", res);
                            uni.setStorageSync('info', res.data[0]);
                            if (this.roleVal == '普通用户') {
                                this.$refs.uToast.show({
                                    message:'登录成功',
                                    type:'success'
                                })
                            };
                            uni.switchTab({
                                url:'/pages/me/me'
                            })
                        } else {
                            this.$refs.uToast.show({
                                message:'无效账号',
                                type:'error'
                            })
                        }
                    }
                });
            }
        }
    }
}
</script>
<style>
    .loginWrap {
        width:100%;
```

```css
        box-sizing:border-box;
        padding:0 40rpx;
    }
    .logo {
        width:30%;
        margin:0 auto;
    }
    .logo image {          width:100%;      }
    .title {          padding-top:60rpx;      }
    .title h3 {
        font-size:24px;
        line-height:40px;
        color:#333333;
    }
    .title h4 {
        font-size:18px;
        color:#888888;
    }
    .formWrap {
        padding-top:20rpx;
    }
    .formItem {
        padding-bottom:15px;
    }
    .formItem .label {
        font-size:14px;
        color:#333333;
    }
    .formItem .inputBox {
        border:1px #C3C3C3 solid;
        margin-top:5rpx;
        border-radius:8px;
    }
    .formItem .inputBox input {
        height:38px;
        line-height:38px;
        text-indent:1em;
    }
    .registTips {
        padding-top:20px;
        text-align:right;
        color:sandybrown;
    }
</style>
```

【代码解析】

单选按钮使用了 uView 中的组件 u-radio，u-radio 需配合 u-radio-group 使用。customStyle 属性可以自定义单选按钮的样式。u-radio-group 双向绑定 roleVal，单选按钮被选中，则值为 label 属性的值。

```
<u-radio-group v-model="roleVal" placement="row">
    <u-radio label="普通用户" name="普通用户" :customStyle="{marginRight:'16px'}" activeColor="rgb(0,194,151)"></u-radio>
```

```
            <u-radio  label="企业用户"  name="企业用户"  activeColor="rgb(0,194,151)">
</u-radio>
        </u-radio-group>
```

3．个人中心页

个人中心页为导航栏中"我的"页面，页面效果如图 7-4 所示。在该页面中点击用户头像，会跳转到登录页。页面完整代码如下。

```
<template>
    <view class="userCenter">
        <view class="userCard">
            <view class=" user-box  ">
                <view class="user-img" @click="changeUser">
                    <u-avatar :src="pic" size="80"></u-avatar>
                </view>
                <view class="user-info">
                    <view class=" ">{{userName}}</view>
                    <view class="user-info-phone">{{phone}}</view>
                </view>
                <view class="sign">
                    <text>签到</text>
                </view>
                <view class="sign-icon">
                    <u-icon name="arrow-right" color="#fff" size="20"></u-icon>
                </view>
            </view>
            <view class="overview">
                <view class="view-item">
                    <view>{{revenue}}</view>
                    <view>累计收益</view>
                </view>
                <view class="view-item">
                    <view>{{orderTimes}}</view>
                    <view>回收次数</view>
                </view>
                <view class="view-item">
                    <view>{{totalScore}}</view>
                    <view>积分</view>
                </view>
            </view>
        </view>
        <view class="overview ">
            <u-cell-group>
                <u-cell icon="star" title="积分记录" isLink :iconStyle="iconStyle">
</u-cell>
                <u-cell icon="photo" title="兑换记录" isLink :iconStyle="iconStyle">
</u-cell>
                <u-cell icon="coupon" title="我的贡献" isLink :iconStyle="iconStyle">
</u-cell>
                <u-cell icon="heart" title="收入记录" isLink :iconStyle="iconStyle">
</u-cell>
                <u-cell icon="setting" title="设置" isLink :iconStyle="iconStyle">
</u-cell>
```

```
                </u-cell-group>
            </view>
        </view>
    </template>
    <script>
        export default {
            data() {
                return {
                    pic:'../../static/avatar.png',
                    show:true,
                    userName:'小甜鱼',
                    phone:'',
                    orderTimes:23,
                    revenue:100,
                    totalScore:100,
                    iconStyle:{
                        padding:'8px 8px'
                    }
                }
            },
            onShow() {
                this.show = uni.getStorageSync('info') != '';
                if(this.show){
                    this.userName = uni.getStorageSync('info').nickName;
                    this.phone = uni.getStorageSync('info').phoneNum.substr(0,7)+'****';
                    console.log(uni.getStorageSync('info').avatar);
                    this.pic= uni.getStorageSync('info').avatar !='' ? this.$baseUrl.baseUrl + uni.getStorageSync('info').avatar :'https://uviewui.com/common/logo.png';
                }else{
                    this.userName ="未登录",
                    this.phone = '';
                    this.orderTimes=0;
                    this.revenue=0;
                    this.totalScore= 0;
                }
            },
            methods:{
                changeUser(){
                    uni.navigateTo({
                        url:'/pages/login/login'
                    })
                }
                //如果有积分记录、回收次数等功能的需求,可以添加相应代码,这里省略此部分代码
            }
        }
    </script>
    <style lang="scss">
    page{   background-color:#ededed; }
    .user-box{
        padding-top:10px;
        color:#fff;
        display:flex;
```

```css
    padding:30 10 20 30;
    align-items:center;
}
.userCenter{
    box-sizing:border-box;
    padding:0 15px;
    background-color:#fff;
    overflow:hidden;
}
.userCard{
    margin:15px auto ;
    background-color:#71d9ca;
    border-radius:10px;
    padding:10px;
    margin-bottom:0px;
}
.user-info{
    flex:1;
    font-size:18px;
    margin-left:15px;
}
.user-info-phone{    font-size:15px;  }
.sign{
    font-size:15px;
    margin-right:15px;
}
.overview{
    display:flex;
    align-items:center;
    padding-top:15px;
}
.view-item{
    flex:1;
    font-size:16px;
    text-align:center;
    color:#fff;
    line-height:24px;
}
.userDetail{
    padding-top:20px;
    padding-bottom:20px;
}
.iconStyle{    padding-top:50px;   }
</style>
```

【代码解析】

页面中的列表使用了 uView 的 u-cel 组件，该组件需要搭配 u-cell-group 使用，用 u-cell-group 实现列表的上下边框。这里 icon 表示列表项左侧的图标，title 表示列表项左侧的标题，默认情况下，列表项右侧有箭头。iconStyle、righticonStyle、titleStyle 属性分别设置列表项左侧的图标、右侧箭头、左侧标题的样式。更多的样式读者可以参考 uView 官网的介绍。

```
<u-cell-group>
    <u-cell icon="star" title="积分记录" isLink  :iconStyle="iconStyle"></u-cell>
</u-cell-group>
```

7.3.5 公司回收页、公司详情页、公司搜索结果页

1. 公司回收页

公司回收页展示回收公司的信息,可以由此页面进入下单页,页面效果如图 7-3 所示。公司信息数据来源地址:http://localhost:3004/companys。点击公司列表项进入公司详情页。另外,有可能公司比较多,在公司信息列表外面加一个 sroll-view 容器,当公司信息在一个页面显示不下时,可以往下滑动。

完整的代码如下。

```
<template>
    <view id="companyRecycle">
        <view class="cpHead">
            <view class="title">
                <image mode="widthFix" src="../../static/app_logo.png"></image>
                <text>公司回收</text>
            </view>
            <view class="search">
                <u-search placeholder="请输入搜索内容" height="50" shape="square" v-model="keyword" @search="search" :showAction="false" ></u-search>
            </view>
        </view>
        <view class="container">
            <scroll-view scroll-y="true" class="companylist">
                <view class="company-item" v-for="item in cpnlist">
                    <view class="item-title">{{item.name}}</view>
                    <view class="item-cont" @click="gotoDetail(item.id)" >
                        <view class="item-img">
                            <image :src="$baseUrl.baseUrl+item.coverUrl"></image>
                        </view>
                        <view class="cont-right">
                            <view class="right-info">{{item.address}}</view>
                            <view class="right-info">营业时间:8:00—17:00</view>
                            <view class="right-info">联系电话:{{item.contact}}</view>
                        </view>
                    </view>
                    <view class="placeOrder">
                        <u-button @click="gotoFillIn(item.id)" type="primary" color="rgb(0,194,151)">立即下单</u-button>
                    </view>
                </view>
            </scroll-view>
        </view>
    </view>
</template>
<script>
    export default {
        data() {
            return {
                cpnlist:[],
```

```
                    keyword:''
                }
        },
        onLoad() {        this.getCompantlist();    },
        methods:{
            getCompantlist() {
                uni.$u.http.get(this.$baseUrl.baseUrl+'/companys', {}).then
(res => {
                    console.log(res);
                    if (res.statusCode == 200) {
                        this.cpnlist = res.data;
                        console.log(this.cpnlist);
                    }
                })
            },
            gotoDetail(id) {
                uni.navigateTo({
                    url:'/childPages/company/detail?id=' + id
                })
            },
            search(e) {
                uni.navigateTo({
                    url:'/childPages/company/searchRes?text=' + e
                })
            },
            gotoFillIn(id,name) {
                uni.navigateTo({
 url:'/childPages/company/fillIn?companyId=' + id+'&companyName='+name
                })
            }
        }
    }
</script>
<style scoped>
    #companyRecycle {        background-color:#F7fafb;    }
    .cpHead {
        box-sizing:border-box;
        padding:20px;
        background-color:rgb(0, 194, 151);
        height:200px;
    }
    .cpHead .title {
        font-size:20px;
        color:#fff;
        display:flex;
        align-items:center;
    }
    .cpHead .title image {    width:60px;    }
    .search {
        width:100%;
        box-sizing:border-box;
        padding:30px 20px 15px 20px;
        background-color:rgb(0, 194, 151);
    }
```

```css
.container {
    width:100%;
    height:calc(100vh - 250px);
    overflow:hidden;
}
.companylist {
    box-sizing:border-box;
    padding:0 20px;
    height:100%;
}
.company-item {
    box-sizing:border-box;
    padding:10px;
    margin:15px 5px 0 5px;
    background-color:#fff;
    border-radius:8px;
    box-shadow:0 1px 8px rgba(14, 197, 156, 0.4);
}
.company-item::after {
    content:'';
    display:block;
    clear:both;
}
.item-title {
    font-size:18px;
    border-bottom:1rpx #e6e6e6 solid;
    line-height:36px;
}
.item-cont {
    display:flex;
    padding:5px 0;
}
.item-img {
    flex:.8;
}
.item-img image {
    width:100%;
    height:90px;
}
.cont-right {
    flex:2;
    margin-left:10px;
    display:flex;
    flex-flow:column;
    justify-content:space-between;
}
.placeOrder {
    width:30%;
    float:right;
}
</style>
```

2. 公司详情页

公司详情页展示单个公司的详细信息,页面效果如图 7-11 所示。其 API 地址为:

http://localhost:3004/companys?id=id。完整的代码如下。

```html
<template>
    <view>
        <view class="cpHead">
            <view class="company-item">
                <view class="item-title">{{companyDetail.name}}</view>
                <view class="item-cont">
                    <view class="item-img">
                        <image :src="$baseUrl.baseUrl + companyDetail.coverUrl"></image>
                    </view>
                    <view class="cont-right">
                        <view class="right-info">地址:{{companyDetail.address}}</view>
                        <view class="right-info">联系电话:{{companyDetail.contact}}</view>
                        <view class="right-info">评价等级:{{companyDetail.score}}</view>
                    </view>
                </view>
            </view>
        </view>
        <view class="introduce">{{companyDetail.introduction}}</view>
        <view class="appointBtn">
            <u-button @click="gotoFillIn(companyDetail.id)" color="rgb(0,194,151)">预约上门回收</u-button>
        </view>
    </view>
</template>
<script>
    export default {
        data() {
            return {
                companyDetail:{}
            }
        },
        onLoad(options) {
            this.getCompanyDtl(options.id)
        },
        methods:{
            getCompanyDtl(id) {
                uni.$u.http.get(this.$baseUrl.baseUrl + '/companys?id='+id,{}).then( res => {
                    console.log(res);
                    if(res.statusCode == 200) {
                        this.companyDetail = res.data[0];
                        console.log(this.companyDetail);
                    }
                })
            },
            gotoFillIn(id) {
                uni.navigateTo({
                    url:'/childPages/company/fillIn?id='+id
                })
```

```
            }
        }
    }
</script>
<style scoped>
    .cpHead{
        box-sizing:border-box;
        padding:20px;
        background-color:rgb(0,194,151);
    }
    .item-title{
        font-size:18px;
        line-height:36px;
        color:#fff;
    }
    .item-cont{
        display:flex;
        padding:5px 0;
        margin-top:10px;
    }
    .item-img{
        flex:.8;
    }
    .item-img image{
        width:100%;
        height:90px;
    }
    .cont-right{
        flex:2;
        margin-left:10px;
        display:flex;
        flex-flow:column;
        justify-content:space-between;
        color:#fff;
    }
    .introduce{
        box-sizing:border-box;
        padding:20px;
    }
    .appointBtn{
        width:calc(100% - 40px);
        position:fixed;
        left:20px;
        bottom:20px;
    }
</style>
```

3. 公司搜索结果页

公司回收页、公司搜索结果页中的搜索框用于输入搜索关键字来搜索名称含有关键字的公司，并将搜索结果显示在公司搜索结果页中，页面效果如图 7-10 所示。其 API 地址为：http://localhsot:3004/companys?name_like=公司名称。这里的公司搜索结果页与分类查询结果页类似，不做过多解释，页面完整代码如下。

```html
<template>
    <view>
        <view class="search">
            <u-search placeholder="请输入搜索内容" shape="square" v-model="keyword" @search="search" actionText="搜索"   :showAction="true"></u-search>
        </view>
        <view class="container">
            <scroll-view scroll-y="true" class="companylist" v-if="cpnlist.length > 0">
                <view class="company-item" v-for="(item,index) in cpnlist" :key="index" @click="gotoDetail(item.id)">
                    <view class="item-title">{{item.name}}</view>
                    <view class="item-cont">
                        <view class="item-img">
                            <image :src="$baseUrl.baseUrl + item.coverUrl"></image>
                        </view>
                        <view class="cont-right">
                            <view class="right-info">{{item.address}}</view>
                            <view class="right-info">营业时间:8:00—17:00</view>
                            <view class="right-info">联系电话:{{item.contact}}</view>
                        </view>
                    </view>
                    <view class="placeOrder">
                        <u-button :customStyle="customStyle" type="primary" color="rgb(0,194,151)">立即下单</u-button>
                    </view>
                </view>
            </scroll-view>
            <view class=" emptyBox" v-else>
                <u-empty text="暂无搜索数据!" mode="data"></u-empty>
            </view>
        </view>
    </view>
</template>
<script>
    export default {
        data() {
            return {
                cpnlist:[],
                keyword:''
            }
        },
        onLoad(options) {
            this.keyword = options.text;
            this.getSearchRes(options.text)
        },
        methods:{
            gotoDetail(id) {
                uni.navigateTo({
                    url:'/childPages/company/detail?id=' + id
                })
            },
            search(e) {
```

```
                    this.getSearchRes(e);
                },
                getSearchRes(str) {
                    uni.$u.http.get(this.$baseUrl.baseUrl + "/companys?name_like="
+ str, {

                    }).then(res => {
                        console.log(res);
                        if (res.statusCode == 200) {
                            this.cpnlist = res.data;
                            console.log(this.cpnlist);
                        }
                    })
                }
            }
        }
</script>
<style scoped>
    .customStyle {
        backgroundColor:'rgb(0,194,151)';
        color:'white';
        height:'32px';
        fontSize:'14px';
        borderRadius:'16px'
    }
    .result {    background-color:#f7fafd;    }
    .search {
        width:100%;
        box-sizing:border-box;
        padding:15px 20px 15px 20px;
        background-color:rgb(0, 194, 151);
    }
    .container {
        width:100%;
        height:600px;
        height:calc(100vh - 130px);
        overflow:hidden;
    }
    .companylist {
        box-sizing:border-box;
        padding:0 15px;
        height:100%;
    }
    .company-item {
        box-sizing:border-box;
        padding:10px;
        margin:15px 5px 0 5px;
        background-color:#fff;
        border-radius:8px;
        box-shadow:0 1px 8px rgba(14, 197, 156, 0.4);
    }
    .company-item::after {
        content:'';
        display:block;
```

```
        clear:both;
    }
    .item-title {
        font-size:18px;
        border-bottom:1rpx #e6e6e6 solid;
        line-height:36px;
    }
    .item-cont {
        display:flex;
        padding:5px 0;
    }
    .item-img {    flex:.8;    }
    .item-img image {
        width:100%;
        height:90px;
    }
    .cont-right {
        flex:2;
        margin-left:10px;
        display:flex;
        flex-flow:column;
        justify-content:space-between;
    }
    .placeOrder {
        width:30%;
        margin-top:5px;
        float:right;
    }
    .emptyBox {
        position:fixed;
        top:50%;
        left:50%;
        transform:translate3d(-50%, -50%, 0);
    }
</style>
```

7.3.6 下单页、订单页、订单详情页

1. 下单页

从公司相关的页面跳转至下单页，跳转时，url 会携带公司的"id"、名称，下单的操作结果就是在订单列表中增加一条记录。页面效果如图 7-9 所示。使用 POST 请求的方式访问 API 地址 http://localhost:3004/companys。由于后端服务器为本地模拟服务器，上传图片不被支持，故此处订单图片为指定图片，上传图片的功能省略。页面完整代码如下。

```
<template>
    <view class="formWrap">
        <u-form :model="form" ref="uForm" label-width="80" labelPosition=
"left" :rules="rules">
            <u-form-item label="物品类别" prop="order.type" borderBottom ref="type"
@click="typeShow = true">
    <u-input v-model="form.order.type" placeholder="请选择物品类别" border="none">
</u-input>
```

```
                    <u-icon slot="right" name="arrow-right"></u-icon>
                    <u-picker :show="typeShow" @confirm="typeCfm" @change="changeHandler" @cancel="typeCancel"
                        :columns="typelist" ref="uPicker" v-model="typeShow"></u-picker>
                </u-form-item>
                <u-form-item label="联系人" prop="order.name" borderBottom>
    <u-input v-model="form.order.name" placeholder="请输入" border="none"></u-input>
                </u-form-item>
                <u-form-item label="联系方式" prop="order.phone" borderBottom>
    <u-input v-model="form.order.phone" placeholder="请输入" border="none"></u-input>
                </u-form-item>
                <u-form-item label="上门地址" prop="order.address" borderBottom>
    <u-input v-model="form.order.address" placeholder="请输入" border="none"></u-input>
                </u-form-item>
    <u-form-item label="上门时间" prop="order.time" borderBottom @click="timeShow = true">
    <u-input v-model="form.order.time" placeholder="请选择上门时间" border="none" disabled disabledColor="#ffffff">        </u-input>
                    <u-icon slot="right" name="arrow-right"></u-icon>
                    <u-action-sheet :show="timeShow" :actions="timelist" title="请选择时间" description="选择上门时间"
                        @close="timeShow = false" @select="timeCfm">
                    </u-action-sheet>
                </u-form-item>
                <u-form-item label="物品重量" prop="order.weight" borderBottom>
    <u-input v-model="form.order.weight" placeholder="请输入" border="none"></u-input>
                </u-form-item>
                <u-form-item label="物品件数" prop="order.number" borderBottom>
    <u-input v-model="form.order.number" placeholder="请输入" border="none"></u-input>
                </u-form-item>
            </u-form>
            <view style="margin:15px auto;">
                <u-upload :fileList="fileList" @afterRead="afterRead" @delete="deletePic" name="2" multiple :maxCount="10">
                </u-upload>
            </view>
            <view class="details">
                <textarea v-model="content" placeholder="请输入其他信息"></textarea>
            </view>
            <view class="appointBtn">
    <u-button type="primary" @click="appointment" color="rgb(0,194,151)">立即预约</u-button>
            </view>
            <u-toast ref="uToast"></u-toast>
        </view>
```

```vue
</template>
<script>
    export default {
        data() {
            return {
                fileList:[],
                latitude:39.909,
                longitude:116.39742,
                form:{
                    order:{
                        type:'',
                        name:'',
                        phone:'',
                        address:'',
                        time:'',
                        weight:'',
                        number:''
                    }
                },
                rules:{
                    'order.type':[{
                        required:true,
                        message:'请选择物品类别',
                        trigger:['change']
                    }],
                    'order.name':[{
                        required:true,
                        message:'请输入联系人',
                        trigger:['blur']
                    }],
                    'order.phone':[{
                        required:true,
                        type:'number',
                        message:'请输入联系方式',
                        trigger:['blur']
                    }],
                    'order.address':[{
                        required:true,
                        message:'请输入地址',
                        trigger:['blur']
                    }],
                    'order.time':[{
                        required:true,
                        message:'请选择上门时间',
                        trigger:['change']
                    }],
                    'order.weight':[{
                        required:true,
                        type:'float',
                        message:'请输入物品重量',
                        trigger:['blur']
                    }],
```

```js
            'order.number':[{
                required:true,
                message:'请输入物品件数',
                type:'integer',
                trigger:['blur']
            }],
        },
        typeShow:false,
        timeShow:false,
        typelist:[
            ['废纸','塑料','金属','纺织品','家电'],
            ['报纸','书籍','纸箱']
        ],
        typeData:[
            ['报纸','书籍','纸箱'],
            ['水瓶','桶','塑料袋','农用地膜','泡沫塑料','塑料编织品'],
            ['铜丝','铁制品','废钢筋','不锈钢','废银锌电池'],
            ['羽绒服','毛衣','棉絮','上衣','裤子','裙子'],
            ['电视机','空调','冰箱','洗衣机','笔记本电脑','微波炉','烤箱']
        ],
        typeId:null,
        unitPrice:null,
        timelist:[{
            name:'8:00—10:00'
        },{
            name:'10:00—12:00'
        },{
            name:'14:00—16:00'
        },{
            name:'16:00—18:00'
        }],
        photoUrl:'',
        content:'',
        companyId:null
        }
    },
    onReady() {
        this.$refs.uForm.setRules(this.rules);
    },
    onLoad(options) {
        let user = uni.getStorageSync('info');
        if (user == '') {
            uni.navigateTo({
                url:'/pages/login/login'
            })
        }
        this.companyId = options.id;
        this.companyName = options.Name;
    },
    methods:{
```

```js
changeHandler(e) {
    const {
        columnIndex,
        value,
        values, // values 为当前变化列的数组内容
        index,
        // 微信小程序无法将 picker 实例传出来，只能通过 ref 操作
        picker = this.$refs.uPicker
    } = e
    // 当第一列值发生变化时，变化第二列（后一列）对应的选项
    if (columnIndex == 0) {
        // picker 为选择器 this 实例，变化第二列对应的选项
        picker.setColumnValues(1, this.typeData[index])
    }
},
typeCancel() {
    this.typeShow = false;
},
typeCfm(e) {
    console.log(e);
    console.log(e.value[1]);
    this.form.order.type = e.value[1];
    this.typeShow = false;
},
timeCfm(e) {
    console.log(e.name);
    this.form.order.time = e.name;
},
appointment() {

    let userId = uni.getStorageSync('info').id;
    this.photoUrl = '';
    this.unitPrice = 2.3;
    this.$refs.uForm.validate().then(valid => {
        console.log("valid=" + valid);
        if (valid) {
            uni.request({
                url:this.$baseUrl.baseUrl + '/orders',
                method:'POST',
                data:{
                    "companyName":this.companyName,
                    "createby":this.form.order.name, //联系人
                    "createTime":this.$u.timeFormat(new Date(), "yyyy-mm-dd hh:MM:ss"),
                    "pickupTime":this.form.order.time, //上门时间
                    "pickupAddress":this.form.order.address, //上门地址
                    "goodNumber":this.form.order.weight, //物品重量
                    "goodUnitPrice":this.unitPrice, //回收物品单价
                    "goodTypeName":this.form.order.type, //回收类型
                    "coverUrl":"/image/order/paper.jpeg",
                    "contact":this.form.order.phone, //联系方式
```

```javascript
                            "remark":this.content, //备注
                            "pkgs":this.form.order.pkgs, //物品包裹数量
                            "state":1,
                            "companyId":this.companyId,
                            "userId":userId
                        },
                        success:(res) => {
                            console.log(res);
                            if (res.statusCode == 201) {
                                this.$refs.uToast.show({
                                    message:'预约成功',
                                    type:'success',
                                    isTab:true

                                })
                                uni.switchTab({
                                    url:'/pages/order/order'
                                })
                            } else {
                                uni.showToast({
                                    message:'预约失败,请重试!',
                                    icon:'error'
                                })
                            }
                        }
                    })
                }
            })
        }
    }
}
</script>
<style scoped>
    .formWrap {
        box-sizing:border-box;
        padding:0 15px 70px 15px;
    }
    .details {
        border:1rpx #E6E6E6 solid;
        border-radius:5px;
    }
    .appointBtn {
        width:calc(100% - 40px);
        position:fixed;
        left:20px;
        bottom:20px;
    }
</style>
```

【代码解析】

该页面使用了 uView 的表单组件 u-form。在<u-form></u-form>标签内通过<u-form-item></u-form-item>标签添加表单项,<u-form-item></u-form-item>标签内除了 input 外,还可以放

置其他的元素。本页面中应用的 u-form 组件的属性及其含义如下。

> - model：表单数据对象。
> - labelWidth：标签的宽度，默认单位为 px。
> - labelPosition：标签的位置，可选值有 top、left，默认值为 left。
> - rules：表单规则，这里对应的是一个规则对象。

本页面中应用的 u-form-item 组件的属性及其含义如下。

> - label：表单项的文本标签。
> - borderBottom：是否显示表单项的底部边框。
> - prop：表单数据对象的属性名，在使用 validate、resetFields 方法的情况下，该属性是必填的。本页面使用了 validate 方法，对表单数据进行有效性验证。下面是联系方式的规则定义，该字段的值只能为数字，不能为空，失去焦点时进行验证，如验证失败，则在下方显示"请输入联系方式"。表单验证的效果如图 7-18 所示。

图 7-18　表单验证效果

```
rules:{
……省略
    'order.phone':[{
        required:true,
        type:'number',
        message:'请输入联系方式',
        trigger:['blur']
    }],
        ……省略
}
```

在 JS 代码中通过下列方法来判断是否通过验证，若 valid 的值为 true，则表示通过验证。

```
this.$refs.uForm.validate().then(valid => {
if (valid) {  }
}
```

本页面中将 input（单行输入框）的边框去掉，然后显示表单项的底部边框。在设置【物品类别】【上门时间】这两项时都会在页面底部弹出选择器或者操作菜单进行选择。效果如图 7-19 和图 7-20 所示。这里【物品类别】选择的是 uView 的 u-picker 组件，实现多级关联选择，【上门时间】选择的是 u-Action-Sheet 操作菜单组件。

2．订单页

订单页展示当前用户的待接单、已接单、已完成、已取消等多种状态的订单信息，如图 7-2 所示。下面只列举当前用户的待接单信息，通过 GET 请求方式访问 API 获取数据。

API 地址为 http://localhost:3004/orders?userId=用户 id &state=1。

图 7-19 多级关联选择

图 7-20 操作菜单选择

其他的订单信息显示方法与此类似,这里不做讲解。以下只列出待接单信息显示的代码,具体代码如下。

```
<template>
    <view class="wrap">
        <view class="u-tab-box">
            <u-subsection :list="list" :current="swiperCurrent" @change=
"sectionChange" activeColor="#00c297"  bgColor="red"
            mode="subsection"></u-subsection>
        </view>
        <swiper class="swiper-box" :current="swiperCurrent" >
            <swiper-item class="swiper-item">
                <scroll-view scroll-y style="height:100%;width:100%;" >
                    <view class="page-box">
    <view class="order" v-for="(item,index) in dataList1"  :key="item.id"
                        @click="gotoDetail(item.id)">
                        <view class="top">
                            <view class="left">
                                <view class="store"> 已等待:{{$u.timeFrom
(getUnixTime(item.createTime+'')),'yyyy-mm-dd')}}</view>
                            </view>
                            <view class="right">待接单</view>
                        </view>
                        <view class="item">
                            <view class="left">
                <image :src="$baseUrl.baseUrl+item.coverUrl" mode="aspectFill">
</image>
                            </view>
                            <view class="content">
    <view class="title u-line-2">回收的类型:{{ item.goodTypeName }}</view>
        <view class="type">重量:{{ item.goodNumber}}kg</view>
        <view class="type">预计收入金额:{{ item.goodNumber*item.goodUnitPrice}}
元</view>
```

```html
                    <view class="type">取货方式:上门收货{{ item.pickupTime }}</view>
                </view>
            </view>
                <view class="bottom">
                    <view class="more">
            <u-icon name="more-dot-fill" color="rgb(203,203,203)"></u-icon>
                    </view>
            <view class="evaluate btn" @tap="cancelOrder(item.id)">取消订单</view>
                        </view>
                </view>
            </view>
        </scroll-view>
    </swiper-item>
    <swiper-item>已接单订单列表</swiper-item>
    <swiper-item>已完成订单列表</swiper-item>
    <swiper-item>已取消订单列表</swiper-item>
    </swiper>
    <u-toast ref='uToast' ></u-toast>/>
    </view>
</template>
<script>
    export default {
        data() {
            return {
                list:['待接单', '已接单', '已完成', "已取消"],
                swiperCurrent:0,
                dataList1:[]
            }
        },
        onLoad() {
            this.getOrders();
        },
        methods:{
            getUnixTime(dateStr) {
                let newstr = dateStr.replace(/-/g, '/');
                let date = new Date(newstr);
                let time_str = date.getTime().toString();
                return time_str.substr(0, 10);
            },
            sectionChange(index) {
                this.swiperCurrent = index;
            },
            getOrders(idx) {
                let user = uni.getStorageSync('info');
                if(user == ''){
                    uni.navigateTo({
                        url:'/pages/me/login'
                    })
                }
                let userId = user.id;
                let url = this.$baseUrl.baseUrl+'/orders?state=1&userId=' + userId;
                uni.request({
                    url:url,
```

```
                    method:'GET',
                    success:(res) => {
                        console.log(res);
                        this.dataList1 = res.data;
                    }
                })
            },
            gotoDetail(id) {
                uni.navigateTo({
                    url:'/childPages/order/waiting?id=' + id
                })
            },
            cancelOrder(id){
                console.log(id);
                uni.request({
                    url:this.$baseUrl.baseUrl+'/orders/'+id,
                    method:'patch',
                    data:{
                        state:0
                    },
                    success:(res) => {
                        if(res.statusCode == 200){
                            this.$refs.uToast.show({
                                type:'success',
                                message:'修改成功,即将刷新页面',
                                duration:3000
                            })
                        };
                        location.reload();//刷新页面
                    }
                })
            }
        }
    }
</script>
<style lang="scss">
    .order {
        width:710rpx;
        background-color:#ffffff;
        margin:20rpx auto;
        border-radius:20rpx;
        box-sizing:border-box;
        padding:20rpx;
        font-size:28rpx;
        border-bottom:1px solid #dddddd;
        .top {
            display:flex;
            justify-content:space-between;
            .left {
                display:flex;
                align-items:center;
                .store {
                    margin:0 10rpx;
                    font-size:32rpx;
```

```
                font-weight:bold;
            }
        }
        .right {
            color:#00c297;
        }
    }
    .item {
        display:flex;
        margin:20rpx 0 0;

        .left {
            margin-right:20rpx;

            image {
                width:200rpx;
                height:200rpx;
                border-radius:10rpx;
            }
        }
        .content {
            .title {
                font-size:28rpx;
                line-height:50rpx;
            }
            .type {
                margin:10rpx 0;
                font-size:24rpx;
                color:$u-tips-color;
            }
            .delivery-time {
                color:#e5d001;
                font-size:24rpx;
            }
        }
        .right {
            margin-left:10rpx;
            padding-top:20rpx;
            text-align:right;

            .decimal {
                font-size:24rpx;
                margin-top:4rpx;
            }
            .number {
                color:$u-tips-color;
                font-size:24rpx;
            }
        }
    }
    .total {
        margin-top:20rpx;
        text-align:right;
        font-size:24rpx;
```

```
            .total-price {
                font-size:32rpx;
            }
        }
        .bottom {
            display:flex;
            margin-top:40rpx;
            padding:0 10rpx;
            justify-content:space-between;
            align-items:center;
            .btn {
                line-height:60rpx;
                width:160rpx;
                border-radius:26rpx;
                border:2rpx solid $u-border-color;
                font-size:26rpx;
                text-align:center;
                color:skyblue;
            }
            .evaluate {
                color:#00c297;
                border-color:#00c297;
            }
        }
    }
    .centre {
        text-align:center;
        margin:200rpx auto;
        font-size:32rpx;
        image {
            width:164rpx;
            height:164rpx;
            border-radius:50%;
            margin-bottom:20rpx;
        }
        .tips {
            font-size:24rpx;
            color:#999999;
            margin-top:20rpx;
        }
        .btn {
            margin:80rpx auto;
            width:200rpx;
            border-radius:32rpx;
            line-height:64rpx;
            color:#ffffff;
            font-size:26rpx;
            background:linear-gradient(270deg, rgba(249, 116, 90, 1) 0%, rgba(255, 158, 1, 1) 100%);
        }
    }
    .wrap {
        display:flex;
        flex-direction:column;
```

```
            height:calc(100vh - var(--window-top));
            width:100%;
        }
        .u-tab-box{
            height:40px;
        }
        .swiper-box {
            flex:1;
        }
        .swiper-item {
            height:100%;
        }
</style>
```

【代码解析】

（1）实现分段显示

分段显示使用了 uView 中的分段器 u-subsection 组件。当点击顶部分段器时，current 属性会发生对应变化，与 current 属性双向绑定的 JS 代码的 data 中的属性 swiperCurrent 会跟着变化。

```
        <u-subsection :list="list" :current="swiperCurrent"></u-subsection>
</template>
```

使用轮播组件显示每段的内容，轮播组件的 current 属性同样绑定属性 swiperCurrent。这样分段器和轮播的当前项可以同步。

（2）显示数据

订单信息可能一屏显示不完，所以需要使用 scroll-view 组件，达到可滚动查看的效果。点击订单项会跳到订单详情页 waiting.vue。这里只实现了待接单的订单详情页，其他状态的订单详情页与之类似，故不做介绍。

（3）取消订单

取消订单就是将订单的状态属性 state 值改为 1。json-server 服务器可使用 PATCH 请求方式，对数据做局部更新。订单取消后，需要刷新订单页，cancelOrder 方法实现了该功能。

3. 订单详情页

在订单页中点击订单信息，会跳转到订单详情页。页面效果如图 7-12 所示。该页面的布局与公司详情页的类似，主要实现取消订单的功能。页面完整的代码如下。

```
<template>
    <view>
        <view class="top">
            <view class="left">        <view class="store">订单详情</view>        </view>
            <view class="right">待接单</view>
        </view>
        <view class="cpHead">
            <view class="company-item">
                <view class="item-title">{{orderDetail.name}}</view>
                <view class="item-cont">
                    <view class="item-img">
        <image :src="$baseUrl.baseUrl +orderDetail.coverUrl" mode="aspectFill"></image>
                    </view>
                    <view class="cont-right">
```

```html
            <view class="right-info">回收类型:{{orderDetail.goodTypeName}}</view>
            <view class="right-info">重量:{{orderDetail.goodNumber}}kg</view>
            <view class="right-info">预计收入金额:{{ orderDetail.goodNumber*orderDetail.goodUnitPrice}}元</view>
                    </view>
                </view>
            </view>
        </view>
        <view class="content">
            <view>订单编号:{{orderId}}</view>
            <view>取货方式:上门取货</view>
            <view>取货时间:{{orderDetail.pickupTime}}</view>
            <view>取货地址:{{orderDetail.pickupAddress}}</view>
            <view>备注:{{orderDetail.remark!=null?orderDetail.remark:'暂无'}}</view>
        </view>
        <view class="introduce">{{orderDetail.introduction}}</view>
        <view class="appointBtn">
            <u-button  @click="cancelOrder(orderDetail.id)"  type="primary" color="rgb(0,194,151)" >取消订单</u-button>
        </view>
        <u-toast ref="uToast"></u-toast>
    </view>
</template>
<script>
    export default {
        data() {
            return {
                orderId:'',
                orderDetail:{}
            }
        },
        onLoad(options) {
            this.getOrderDtl(options.id)
        },
        methods:{
            getUnixTime(dateStr) {
                let date = new Date(dateStr);
                let time_str = date.getTime().toString();
                return time_str.substr(0, 10);
            },
            getOrderDtl(id) {
                uni.$u.http.get(this.$baseUrl.baseUrl + '/orders?id='+ id ).then(res => {
                    if (res.statusCode == 200) {
                        console.log(res);
                        this.orderDetail = res.data[0];
                        let str = this.getUnixTime(res.data[0].createTime);
                        this.orderId = str.toString() + id;
                    }
                })
            },
            cancelOrder(id){
```

```
                    console.log(id);
                    uni.request({
                        url:this.$baseUrl.baseUrl+'/orders/'+id,
                        method:'patch',
                        data:{
                            state:0
                        },
                        success:(res) => {
                            if(res.statusCode == 200){
                                this.$refs.uToast.show({
                                    type:'success',
                                    message:'修改成功,即将刷新页面',
                                    duration:3000
                                });
                                uni.navigateBack();//回到上一级页面
                            }else{
                                this.$refs.show({
                                    message:'操作失败!!',
                                    type:'error',

                                })
                            }

                        }
                    })
                }
            }
        }
</script>
<style lang="scss" scoped>
    .top {
        display:flex;
        justify-content:space-between;
        margin-bottom:20rpx;
        margin-left:20rpx;
        margin-right:20rpx;
        .left {
            display:flex;
            align-items:center;
            border-left:4rpx solid rgb(0, 194, 151);
            .store {
                margin:0 10rpx;
                font-size:32rpx;
                font-weight:bold;
            }
        }
        .right {
            color:rgb(255, 100, 0);
            font-size:32rpx;
            font-weight:bold;
        }
    }
    .cpHead {
        box-sizing:border-box;
```

```css
        padding:20px;
        background-color:rgb(0, 194, 151);
        margin:0 20rpx;
        border-radius:8rpx;
    }
    .item-title {
        font-size:18px;
        line-height:36px;
        color:#fff;
    }
    .item-cont {
        display:flex;
        padding:5px 0;
        margin-top:10px;
    }
    .item-img {    flex:.8;    }
    .item-img image {
        width:100%;
        height:90px;
    }
    .cont-right {
        flex:2;
        margin-left:10px;
        display:flex;
        flex-flow:column;
        justify-content:space-between;
        color:#fff;
    }
    .introduce {
        box-sizing:border-box;
        padding:20px;
    }
    .appointBtn {
        width:calc(100% - 40px);
        position:fixed;
        left:20px;
        bottom:20px;
    }
    .content {
        font-size:15px;
        line-height:30px;
        margin-top:20px;
        margin-left:20px;
    }
</style>
```

本章小结

本章基于 json-server 工具模拟后端数据接口，结合 uView 2.0 组件开发了一个智慧环保项目。json-server 是一个在前端本地运行，可以存储 JSON 数据的服务器。uView 是一个优秀的 uni-app 生态框架，其全面的组件和便捷的工具可以帮助用户快速开发移动应用项目。智慧环保项目功能完善、链接正常，能培养读者的综合实践能力。

项目实战

由于篇幅的限制，智慧环保项目中的积分商城、附近回收、设置、公司端等功能尚未实现，请继续完善项目。表 7-3 列出了该项目的部分扩展功能，仅供参考。

表 7-3 项目的部分扩展功能

模块	功能	功能描述
积分商城	商品列表	显示可兑换的商品列表
附近回收	回收机列表	显示当前城市所有回收机及其在地图上的位置
	回收机详情	显示回收机详细信息
	回收机回收	显示回收机业务流程
设置	修改密码	需输入旧密码、新密码、确认密码
	更换手机号	需输入新手机号、验证码或旧密码
公司端	登录	登录、切换用户、退出登录
	处理订单	接单、取消订单、查看订单详情
	个人中心	企业账号的相关信息

拓展实训项目

时代楷模是由中共中央宣传部集中组织宣传的全国重大先进典型。时代楷模充分体现"爱国、敬业、诚信、友善"的社会主义核心价值观，充分体现中华传统美德，是具有很强先进性、代表性、时代性和典型性的先进人物。时代楷模事迹厚重感人、道德情操高尚、影响广泛深远。根据时代楷模的职业身份，以中共中央宣传部和有关部门名义发布时代楷模。在中央电视台设立了"时代楷模发布厅"。

随着经济快速发展，计算机的普及率越来越高，互联网用户数量逐年增多，在多元网络文化环境中，年轻人容易被负面文化影响，误入歧途。时代楷模 App 运用新信息技术，整合多方资源，让更多的年轻人通过该 App 看要闻、学新思想、明历史、长知识、知晓时政，紧跟党的步伐，不断武装思想。它主要包括以下功能模块。

1. 楷模公告：在首页以幻灯片轮播图展示时代楷模精神、往期公告列表。
2. 楷模列表：包括楷模介绍、楷模事迹视频、致敬内容、评论等。
3. 楷模故事：包括楷模事迹介绍、评论等，事迹介绍包括文章、视频等资源。
4. 学习心得：包括学习笔记、学习感言、学习历史等。
5. 公益活动：包括活动发起、活动展示、活动报名等，活动内容包括文章、视频等资源。
6. 身边的楷模：将身边的符合时代楷模定义的事迹发布到平台，传递正能量，事迹内容包括文章、照片、视频等资源。

参考文献

[1] 刘刚. 微信小程序开发项目教程（慕课版）[M]. 北京:人民邮电出版社,2021.

[2] 段仕浩,黄伟,赵朝辉. Android移动应用开发案例教程（慕课版）[M]. 北京:人民邮电出版社,2022.

[3] 袁龙. Vue.js核心技术解析与uni-app跨平台实战开发[M]. 北京:清华大学出版社,2021.

[4] 李杰. uni-app多端跨平台开发从入门到企业级实战[M]. 北京:中国水利水电出版社,2022.

[5] 虞芬,张扬,靳红霞. 微信小程序开发与实战（微课版）[M]. 北京:人民邮电出版社,2022.

[6] 刘兵. 鸿蒙应用开发从入门到实战[M]. 北京:中国水利水电出版社,2022.